Molecular Mechanisms of Adult Stem Cell Aging

Else Kröner-Fresenius Symposia

Vol. 1

Series Editor

S. Pahernik Heidelberg

Molecular Mechanisms of Adult Stem Cell Aging

Volume Editor

Karl Lenhard Rudolph Ulm

21 figures, 19 in color, 2010

KARGER

Basel · Freiburg · Paris · London · New York · Bangalore · Bangkok · Shanghai · Singapore · Tokyo · Sydney

Karl Lenhard Rudolph
University of Ulm
Institute of Molecular Medicine and
Max Planck Research Group on Stem Cell Aging
Life Science Building, N27, Room 4015
James-Frank-Ring
DE–89081 Ulm

This book is sponsored by the Else Kröner-Fresenius-Stiftung.

Library of Congress Cataloging-in-Publication Data

Else Kröner-Fresenius Symposium on the Molecular Mechanisms of Adult Stem Cell Aging (1st : 2009 : Günzburg, Germany)
Molecular mechanisms of adult stem cell aging / volume editor, Karl Lenhard Rudolph.
p. ; cm. -- (Else Kröner-Fresenius symposia, ISSN 1663-0114 ; v. 1)
Includes bibliographical references and index.
ISBN 978-3-8055-9243-7 (hard cover : alk. paper)
1. Stem cells--Congresses. 2. Cells--Aging--Congresses. 3. Aging--Molecular aspects--Congresses. I. Rudolph, K. Lenhard. II. Title. III. Series: Else Kröner-Fresenius symposia, v. 1. 1663-0114 ;
[DNLM: 1. Adult Stem Cells--physiology--Congresses. 2. Cell Aging--genetics--Congresses. 3. DNA Damage--physiology--Congresses. 4. Telomere--genetics--Congresses. QU 325 E49m 2010]
QH588.S83E447 2009
616'.02774--dc22

2010007472

Bibliographic Indices. This publication is listed in bibliographic services, including Current Contents®.

Cover art by Maren Blaschke

www.karger.com
Printed in Switzerland on acid-free and non-aging paper (ISO 9706) by Reinhardt Druck, Basel
ISSN 1663–0114
ISBN 978–3–8055–9243–7
eISBN 978–3–8055–9244–4

Contents

Preface

This new book series features the proceedings of the Else Kröner-Fresenius Symposia, which is intended to cover clinically relevant topics at the forefront of biomedical research. They give experts in new or evolving fields of biomedicine the opportunity to critically analyze the most recent findings and outline future research strategies. Today's research is characterized by the accelerated generation of biological and medical data, the increasingly interdisciplinary nature of scientific approaches as well as efforts to integrate individual findings into complex system networks. These developments emphasize the need for new forums of discussion.

The research discussed at the first Else Kröner-Fresenius Symposium focuses on stem cell aging at the cellular and molecular levels. The understanding of stem cell aging in the context of other tissues and malignant tumors is highly relevant for current developments in biomedicine.

Decreasing capacity for tissue renewal is a fundamental feature of the physiological process of aging. Dr. Karl Lenhard Rudolph's research has contributed to our understanding of the molecular mechanisms by which the complex balance of tissue integrity is gradually altered in the process of tissue aging. He has shown that inhibition of telomere dysfunction can modulate the process of cellular senescence. The physiology, modulation and pathology of stem cell systems discussed at the first Else Kröner-Fresenius Symposium promise to eventually open new ways of restoring organ function lost by disease or physical trauma. A wide variety of organ systems, pathophysiological settings and experimental models were discussed at the meeting, integrating knowledge from diverse fields of research.

The Else Kröner-Fresenius-Stiftung thanks Dr. Rudolph for his inspiring scientific work and personal input for organizing together, with his young team, the first Else Kröner-Fresenius Symposium.

The Else Kröner-Fresenius-Stiftung

In 1983, Else Kröner (1925–1988) founded the Else Kröner-Fresenius-Stiftung, a non-profit-making foundation dedicated to promoting medical science and providing humanitarian aid. After her death in 1988, nearly her entire estate was transferred to the foundation for the pursuit of these aims. The symposia are published as part of the foundation's commitment to the advancement of medical research and treatment.

In 1946, after the death of Dr. Eduard Fresenius, who founded the pharmaceutical company Fresenius in 1912, Else Kröner inherited the company. She also inherited the historic Hirsch Apotheke, which had been founded in Frankfurt am Main in the 15th century. Else Kröner decided to take responsibility for the Fresenius company, although it was in debt and all but 30 of the original 400 employees had to be laid off. She joined with her husband, Hans Kröner, to save and expand the company. While the company's mere survival had marked all of its activity to that point, entrepreneurial decisions in the 1950s ensured its successful development. Decades of growth followed, leading to an internationally competitive enterprise and market leader in many areas of health care. Else Kröner led the company until 1981. After the transformation of Fresenius into a stock company, she remained Chairman of the Supervisory Board until her death in 1988. The Fresenius Group is currently the leading global provider of dialysis and artificial nutrition as well as being a major private hospital management company in Germany.

Sascha Pahernik, Heidelberg
Member of the Scientific Commitee of the Else Kröner-Fresenius-Stiftung
Series Editor

Rudolph KL (ed): Molecular Mechanisms of Adult Stem Cell Aging.
Else Kröner-Fresenius Symp. Basel, Karger, 2010, vol 1, pp 1–2

Introduction

K. Lenhard Rudolph

Institute of Molecular Medicine and Max Planck Research Group on Stem Cell Aging, University of Ulm, Ulm, Germany

There is growing evidence that adult stem cells age. This process can result in alterations in stem cell number and function, leading to distinct phenotypic outcomes in different organ systems.

The molecular causes of stem cell aging remain to be defined. There is evidence that stem cell aging can involve cell intrinsic as well as extrinsic alterations that affect the stem cell niche or the macro-environment. In various organ systems, stem cells represent the most long-living population of cells retaining a capacity to proliferate and differentiate. The aging of adult stem cells likely plays a key role in the decline of organ maintenance and regenerative reserve during aging and at the end stage of chronic diseases. In addition, it may contribute to stem cell transformation and carcinogenesis.

Given the pivotal role of adult stem cells in aging, regeneration and cancer, the Else Kröner-Fresenius Stiftung decided to initiate a symposium focusing on adult stem cell aging. The symposium, in May 2009, brought together leading experts in the field in an atmosphere that fostered scientific exchange and open discussion on new models and future concepts. The meeting significantly contributed to scientific advancement in this emerging field of research, which should ultimately lead to the development of new therapies in aging, regeneration and cancer. This book represents a meeting report summarizing the current knowledge in the field of stem cell aging and reviewing the emerging topics discussed at the Else Kröner-Fresenius Symposium on the Molecular Mechanisms of Adult Stem Cell Aging.

Prof. Dr. med. K.L. Rudolph
Life Science Building, N27, Room 4015
James-Franck-Ring
DE–89081 Ulm (Germany)
Tel. +49 731 50 361 00, Fax +49 731 50 361 02
E-Mail lenhard.rudolph@uni-ulm.de

Chapter 1

Rudolph KL (ed): Molecular Mechanisms of Adult Stem Cell Aging.
Else Kröner-Fresenius Symp. Basel, Karger, 2010, vol 1, pp 3–16

Speakers at the Symposium

Daniel Hartmann

Institute of Molecular Medicine and Max Planck Research Group on Stem Cell Aging, University of Ulm, Ulm, and Department of Surgery, Technical University Munich, Munich, Germany

Thomas Braun, MD

T.B. is director of the Department of Cardiac Development and Remodeling at the Max Planck Institute for Heart and Lung Research in Bad Nauheim (Germany) (www.mpi-bn.mpg.de). After finishing his medical studies in Göttingen and his medical thesis at the Institute of Human Genetics in Hamburg, he worked as a postdoc at the Department of Toxicology, University of Hamburg, in Hamburg (Germany), the Institute of Virology in Oxford (UK), the Medical Research Council in Cambridge (UK), the Whitehead Institute for Biomedical Research in Cambridge (USA) and as a group leader at the Department of Cellular and Molecular Biology, Braunschweig University of Technology, in Braunschweig (Germany). In 1993, he completed his PhD in cellular biochemistry and the German 'habilitation'. He has been associate professor at the Institute of Medical Radiology and Cell Research, Würzburg (Germany) and full professor and director of the Institute of Physiological Chemistry at the University of Halle-Wittenberg where he also served as vice dean for research in the Medical Faculty. Since 2004, he has been a scientific member and director at the Max Planck Institute in Bad Nauheim and full professor at the Department of Internal Medicine at the University of Giessen (Germany). He is an elected member of the German Academy of Natural Scientists, Leopoldina and St. Cross College, Oxford (UK).

T.B.'s research has two main focuses. The first is processes that lead to proliferation of organ-typical precursor cells and their coordinated differentiation during organ development and regeneration. The second is the development of pre-clinical models, which can be used to enable, improve and accelerate tissue regeneration, particularly in the heart. He has made several seminal contributions to the understanding of muscle development and regeneration that have had a major impact in the field. Key findings were the identification of the myogenic factor Myf-5, the analysis of the biological function of Myf-5, Myf-6 and MyoD in vivo, the discovery of the homeobox protein Lbx1 as a regulator of limb muscle precursor cell migration, the analysis of the role of FGFs for muscle cell migration and regeneration, the identification of Pax7 as a crucial regulator of muscle satellite cell survival and maintenance, the identification of new muscle-specific transcriptional co-regulators and many other discoveries, including the recent identification of miRNAs as regulators of phenotype modulation of smooth muscle cells, which has a major impact on the understanding of the pathogenesis of arteriosclerosis.

Tao Cheng, MD

T.C. received his medical degrees and clinical training (internal medicine and hematology) from the Secondary Military Medical University, Shanghai (China). He did his postdoctoral fellowship (hematopoiesis and stem cell biology) at the Massachusetts General Hospital and Harvard Medical School with Dr. David Scadden. He is now an associate professor with tenure at the Department of Radiation Oncology and the director of stem cell biology and co-leader of the Cancer Stem Cell Program at the University of Pittsburgh Cancer Institute (http://bmg.mgb.pitt.edu). Prior to his current tenure, he was assistant professor of medicine at Harvard Medical School. He also holds an adjunct professorship at the Chinese Academy of Medical Sciences and Peking Union Medical College and is leading a national laboratory on experimental hematology in China (http://english.bjhb.gov.cn). He has received many awards, including the Young Faculty Scholar Award from the American Society of Hematology in 2002, a Chang-Jiang Scholarship from the Ministry of Education of China in 2007, and the Scholar Award from the Leukemia and Lymphoma Society of the USA in 2008.

T.C.'s research mainly concerns cell cycle control of stem cells and stem cell response to injury or disease. His focus is on (1) the roles of cell cycle regulators (mainly CDK inhibitors) in stem cell self-renewal, (2) the distinction between molecular mechanisms in leukemic stem cells versus normal hematopoietic stem cells, and (3) stem cell protection in transplant recipients or under pathological conditions.

James DeGregori, PhD

J.D. is a professor in the Department of Biochemistry and Molecular Genetics and director of the Molecular Biology Program at the University of Colorado Denver School of Medicine (www.uchsc.edu/molbio). He received his PhD from the Massachusetts Institute of Technology in 1993 under the mentorship of H. Earl Ruley. From 1993 to 1997 he was a postdoctoral fellow with Joseph Nevins at Duke University Medical Center. His honors include being named a V Foundation Scholar and a Leukemia and Lymphoma Society Scholar.

Studies to better understand the conditions that foster the initiation of leukemias and lymphomas are currently a major thrust of the DeGregori lab. The lab has developed an evolution-based model for cancer development, 'Adaptive Oncogenesis', and is currently exploring how reduced progenitor cell fitness resulting from carcinogen exposure, irradiation, inadequate diet or aging can select for adaptive oncogenic events and thereby promote the expansion and fixation of oncogenically initiated cells. In particular, current studies are focused on how the reduced fitness of lymphoid progenitor cells in aged individuals can increase the selection for oncogenic events such as Bcr-Abl, leading to increased leukemogenesis.

Gerald de Haan, PhD

G.d.H. is a professor of molecular stem cell biology at the Department of Cell Biology, University Medical Center Groningen in the Netherlands (www.rug.nl/umcg). He received his PhD in 1995 and was a postdoctoral fellow in the lab of Gary Van Zant at the University of Kentucky until 1998. He was awarded a fellowship by the Royal Netherlands Academy of Arts and Sciences to establish his own lab in Groningen, and received a VICI grant from the Netherlands Organization for Scientific Research.

G.d.H.'s research interests relate to the molecular understanding of self renewal of hematopoietic stem cells. These include the identification of genes that regulate self renewal, hematopoietic stem cell expansion, development of leukemia, and studies on hematopoietic stem cell aging. In earlier studies, G.d.H.'s group was able to show how hematopoietic stem cell turnover is correlated with mouse lifespan. Genetic studies have identified genomic loci that control these parameters, and more recently genome-wide expression studies have resulted in the construction of gene networks that underlie stem cell turnover and functioning. In addition, his lab studies the involvement of epigenetic modifications during hematopoietic stem cell aging.

Ronald A. DePinho, MD

R.A.D. is director of the Belfer Institute Center for Applied Cancer Science at the Dana-Farber Cancer Institute, and is professor of medicine and genetics at Harvard Medical School; he is an American Cancer Society research professor, and member of the Institute of Medicine of the National Academies (http://research4.dfci.harvard.edu/DePinho). He founded and directs the Belfer Institute, which is a novel integrated program for cancer drug discovery and development. R.A.D. serves on numerous advisory boards in the public and private sector and is founder, director and advisor of several biotechnology companies and advises many pharmaceutical companies as well as the US government on various national efforts such as the human cancer genome project. He has received numerous awards including the Melini Award for Biomedical Excellence, the ACSI Stanley J. Korsmeyer Award, the AACR Clowes Award, the Harvey Society Lectureship, the Helsinki Medal, and the Albert Szent-Gyorgyi Prize.

The DePinho laboratory has utilized molecular, cellular and organismal approaches to dissect complex human disorders including cancer and aging. R.A.D. illuminated how the fundamental confluence of telomere dysfunction, impaired DNA damage signaling, and age- or disease-accelerated epithelial renewal conspire to drive the benign to malignant transition in carcinomas, the most common cancers, as well as in aging-related degenerative diseases. These studies revealed how such cancers acquire complex translocations and recurrent amplifications/deletions needed to drive malignancy. Building on these discoveries, he created many engineered cancer models, including those with telomere dysfunction with which he defined fundamental mechanisms underlying cancer's hallmark feature of genome instability and harnessed such systems in human cancer gene discovery. He defined the extent to which telomeres influence the normal aging process and metabolism and established the essentiality of telomeres and FoxOs in stem cell homeostasis in the aged. He provided the first genetic evidence that premature aging and degenerative phenotypes of Werner and Ataxia-Telangiectasia syndromes are driven by the impact of these deficiencies on telomere maintenance. Notably, he showed that end-stage liver cirrhosis, a leading cause of death worldwide, is precipitated by telomere depletion. R.A.D. also made fundamental contributions on many fronts, including the use of inducible cancer models establishing the concept of tumor maintenance and their use in drug response and diagnostics, the genetic analysis of tumor suppressors, and the discovery of co-activated receptor tyrosine kinases in all solid tumors – a paradigm that now guides receptor tyrosine kinase inhibitor combination trials.

Kenneth Dorshkind, PhD

K.D., professor and vice chair for research in the Department of Pathology and Laboratory Medicine at the David Geffen School of Medicine at the University of California, Los Angeles, earned his doctorate in biological structure at the University of Washington, Seattle in 1980 and then completed a postdoctoral fellowship at the Ontario Cancer Institute in Toronto. He serves as director of the Hematopoietic Malignancies Program in the Jonsson Comprehensive Cancer Center (www.cancer.ucla.edu) and as academic associate director of the UCLA Broad Stem Cell Research Center (www.stemcell.ucla.edu). Work in the Dorshkind laboratory is supported by multiple grants from the National Institutes of Health.

Research in the Dorshkind laboratory is focused on the analysis of lymphocyte development during embryogenesis and senescence. One aim is to compare and contrast B cell progenitors that emerge during fetal life with those that are generated in postnatal bone marrow. Emerging data from his laboratory indicate that the first wave of B cell development in the fetus generates B-1 B cell progenitors, which are effectors of innate immunity. The other main goal of the laboratory is to determine the basis for declines in lymphocyte development with age. It is known that lymphoid progenitors exhibit severe growth defects during aging while myelopoiesis is relatively unperturbed. Recent data generated in his laboratory have demonstrated that the preferential expression of the p16Ink4a and Arf tumor suppressor proteins in aged lymphoid progenitors contributes to their reduced growth and survival and makes the aged progenitors refractory to transformation. Down-regulation of p16Ink4a and Arf reversed the senescent phenotype but restored susceptibility to transformation. These data have provided a molecular explanation for lymphoid lineage aging and support the hypothesis that aging and cancer resistance are linked processes.

Toren Finkel, MD, PhD

T.F. is currently the chief of the Translational Medicine Branch of the National Heart Lung and Blood Institute in Bethesda, Md. (USA). He received his undergraduate degree in physics and his MD and PhD degree from Harvard Medical School in 1986. Following a residency in internal medicine at the Massachusetts General Hospital, he completed a fellowship in cardiology at Johns Hopkins Medical School. In 1993, he accepted a position within the Intramural Research Program of the National Institutes of Health in Bethesda. In 2001, he became the chief of the Cardiology Branch and in 2007, he became chief of the newly formed Translational Medicine Branch within the NHLBI (http://dir.nhlbi.nih.gov).

T.F.'s current research interests include (1) the role of reactive oxygen species in aging and the contribution of stem/progenitor cell dysfunction in age-related diseases, (2) the role of mammalian sirtuins in both aging and metabolism, and (3) the mTOR pathway in aging and energy signal transduction. His work is supported by National Institutes of Health intramural funds and from a Senior Scholar award from the Ellison Medical Foundation.

Hartmut Geiger, PhD

H.G. holds the leadership position of the KFO142: 'Molecular and Cellular Aging – From Mechanisms to Clinical Perspective' in the Division of Dermatology and Allergic Diseases at University of Ulm (www.uni-ulm.de/klinik/derma/e/index_e.html), and an adjunct associate professor position in the Division of Experimental Hematology and Cancer Biology at Cincinnati Children's Hospital Medical Center (www.cincinnatichildrens.org). After obtaining his PhD at the Max Planck Institute of Immunobiology in Freiburg (Germany), H.G. moved to the laboratory of Gary Van Zant at the University of Kentucky in Lexington, Ky. (USA) to study the genetic regulation of hematopoietic stem cells. He was subsequently appointed assistant professor and in 2008 associate professor at the Division of Experimental Hematology at Cincinnati Children's Hospital Medical Center.

H.G. was appointed to his current leadership position of the KFO142 in 2008. His research focuses on hematopoietic stem cell biology, with a special emphasis on molecular pathways of stem cell aging and alterations of stem cell niche interaction upon aging.

Margaret A. Goodell, PhD

M.A.G. received her doctorate from the University of Cambridge in England and underwent postdoctoral training at MIT and Harvard Medical School. M.A.G. has been on the faculty at Baylor College of Medicine in Houston, Tex. (USA) (www.bcm.edu/db) since 1997, and is the director of the college's Stem Cells and Regenerative Medicine Center (www.bcm.edu/star). She is a member of the Center for Cell and Gene Therapy, and the Departments of Pediatrics, Molecular and Human Genetics, and Immunology. She has served on the board of the International Society for Stem Cell Research (2005–2008) and the International Society for Experimental Hematology (2009–2012). She serves on several editorial boards including *PLoS Biology*, *Cell Stem Cell*, and *Blood*, and as a reviewer for multiple journals and grant-awarding agencies. She received the DeBakey Award for Excellence in Research at Baylor College in 2004 and the Stohlman Scholar Award from the Leukemia and Lymphoma Society in 2006. In 2007, she was awarded the Vivian L. Smith Chair in Regenerative Medicine at Baylor College.

M.A.G. developed the innovative side population method for the isolation of hematopoietic stem cells from mouse bone marrow, which is now widely used and has been applied to identify candidate stem cells from multiple tissues and species. She now directs a laboratory of about 20 students and postdoctoral fellows focusing on the fundamental mechanisms that regulate hematopoietic stem cells. She has shown that aging has a broad impact on the function and gene expression of mouse stem cells, and is currently studying the specific roles of a variety of genes in stem cell aging.

D. Leanne Jones, PhD

D.L.J. is in the Laboratory of Genetics at the Salk Institute for Biological Studies in La Jolla, Calif. (USA) (www.salk.edu). After completing her PhD in microbiology and molecular genetics at Harvard Medical School with Dr. Karl Münger, she engaged in postdoctoral studies with Philip Ingham at the Medical Research Council Centre for Development and Biomedical Genetics in Sheffield (UK), and then with Margaret Fuller in the Department of Developmental Biology at Stanford University School of Medicine in Palo Alto, Calif. (USA). She started her own research group at the Salk Institute in 2004. D.L.J. has received several research awards including ones from the Ellison Medical Foundation, American Cancer Society, and the California Institute for Regenerative Medicine.

D.L.J.'s research focuses on the molecular mechanisms underlying the manner in which aging affects stem cells, the stem cell environment (niche) and the relationship between the two. Using *Drosophila* spermatogenesis as a model system for studying the aging of adult stem cells, her lab has found that significant changes to the stem cell niche occur, which are accompanied by a concomitant loss of stem cells. An aging-related decline in the expression of key stem cell self-renewal factors normally produced by supporting niche cells revealed that stem cell niches are highly dynamic, rather than static structures. These studies suggest genetic programs are in place to regulate maintenance of a functional stem cell niche over time. Therapeutic strategies that manipulate the size and activity of stem cell niches will complement stem cell transplantation in regenerative medicine and the treatment of cancer.

Zhenyu Ju, MD, PhD

Z.J. heads the Max Planck Partner Research Group on Stem Cell Aging at the Institute of Laboratory Animal Sciences, Chinese Academy of Medical Sciences (www.cnilas.org). After completing his medical studies at Shandong Medical College in Jinan (China) and the China Medical University in Shenyang (China), he engaged in a PhD and postdoctoral training with Lenhard Rudolph at Hannover Medical School (Germany). He started his own research group in Beijing (China) in 2007 in continuous collaboration with Lenhard Rudolph.

Z.J.'s research focuses on the molecular mechanisms underlying the aging process of stem cells. He has carried out projects on the functional consequences of telomere shortening in vivo. His main contributions include the demonstration that (1) telomere dysfunction can induce cell intrinsic checkpoints limiting stem cell maintenance and function, (2) deletion of specific DNA damage checkpoints can improve stem cell function, organ maintenance, and the lifespan of telomere dysfunctional mice without increasing cancer formation, and (3) telomere dysfunction can limit hematopoietic stem cell function through age-dependent environmental alterations.

Sean J. Morrison, PhD

S.J.M. heads the University of Michigan Center for Stem Cell Biology (www.med.umich.edu/cdb). He is a Henry Sewall professor in medicine and an investigator at the Howard Hughes Medical Institute. He obtained his BSc in biology and chemistry from Dalhousie University, then completed a PhD in immunology at Stanford University and a postdoctoral fellowship in neurobiology at Caltech. Since 1999, S.J.M. has been at the University of Michigan, where his laboratory studies the mechanisms that regulate stem cell self-renewal and stem cell aging, as well as the role these mechanisms play in cancer. S.J.M. was a Searle Scholar (2000–2003), was named to Technology Review Magazine's list of 100 young innovators (2002), received the Presidential Early Career Award for Scientists and Engineers (2003), the International Society for Hematology and Stem Cell's McCulloch and Till Award (2007) and the American Association of Anatomists Harland Mossman Award (2008).

The Morrison laboratory studies the cellular and molecular mechanisms that regulate stem cell function in the nervous and hematopoietic systems and the role that these mechanisms play in cancer. In particular, they study the mechanisms that regulate stem cell self-renewal, and how changes in self-renewal pathways can contribute to cancer and stem cell aging. The Morrison laboratory has shown that networks of proto-oncogenes and tumor suppressors, originally discovered for their role in cancer, also regulate stem cell self-renewal. Indeed, proto-oncogenes and tumor suppressors probably evolved to regulate normal stem/progenitor cell function, and their role in cancer reflects the ability of cancer cells to hijack these mechanisms. These mechanisms also change with age, allowing stem cell function to be regulated in a manner that reflects the changing growth and regenerative demands of tissues, while also guarding against the increased risk of cancer during aging.

Hiromitsu Nakauchi, MD, PhD

H.N. is the director of the Division of Stem Cell Therapy and the Center for Stem Cell and Regenerative Medicine at the Institute of Medical Science at the University of Tokyo (www.ims.u-tokyo.ac.jp/imsut/en). He obtained his MD degree from the Yokohama City University School of Medicine and his PhD in immunology from the University of Tokyo Graduate School of Medicine. From 1983 to 1985, he was a postdoctoral fellow in immunogenetics and molecular biology at the Department of Genetics at Stanford University School of Medicine with Prof. Leonard A. Herzenberg. After coming back to Japan, he started working on hematopoietic stem cells in the Laboratory of Molecular Regulation of Aging (Frontier Research Program), and later as a team leader in the Laboratory of Cell Growth and Differentiation at the Institute of Physical and Chemical Research (RIKEN). In 1994, he was appointed to a professorship in immunology at the University of Tsukuba. Since April 2002, he has been a professor of stem cell therapy at the University of Tokyo. In 2008, he became the first director of the newly established Center for Stem Cell Biology and Regenerative Medicine at the Institute of Medical Science.

During his post-doc period H.N. worked on the isolation of CD8 genes at the laboratory of Leonard A. Herzenberg at Stanford University. During his time at the University of Tsukuba he was able to determine the phenotype and purify mouse hematopoietic stem cells by discovering the CD34 surface marker. In addition, he established an in vivo clonal analysis system using single cell transplantation. Goals of his current work are to clarify the mechanism of stem cell self-renewal and to expand the field of cell/gene therapy and regenerative medicine. H.N. has received several research awards, including the Yokohama Medical Award in 1996 and the Erwin von Baelz Prize in 2004.

Andre Nussenzweig, PhD

A.N. received his PhD from the Department of Physics at Yale University. He obtained postdoctoral training at the École Normale Supérieure in Paris and at the Memorial Sloan-Kettering Cancer Center before joining the Experimental Immunology Branch at the Center for Cancer Research of the National Cancer Institute in Bethesda, Md. (USA) where the focus of his research is on the mechanisms by which cells monitor and repair DNA double-strand breaks (http://ccr.cancer.gov).

Recently, the Nussenzweig laboratory was able to show that components of the nonhomologous end-joining pathway, which repairs breaks during V(D)J and class-switch recombination, are caretaker genes that maintain the integrity of the genome. Mice lacking both a nonhomologous end-joining protein (Ku80 or Ku70) and the p53 tumor suppressor protein develop pro-B cell lymphomas at an early age that result from a specific set of chromosomal translocations and gene amplifications involving IgH and c-myc, reminiscent of Burkitt's lymphoma. Currently, the group is studying the mechanisms by which oncogenic translocations are generated in the absence of nonhomologous end joining and is characterizing proteins that mediate DNA damage detection and signaling functions during V(D)J and class-switch recombination using molecular approaches that include targeted gene inactivation and transgenic mice. Overall, the goal of the group is to elucidate the mechanisms by which DNA damage detection, signaling and repair are integrated during lymphocyte development and activation.

Thomas A. Rando, MD, PhD

T.A.R. is professor of neurology and neurological sciences at Stanford University School of Medicine (www.stanford.edu/~casco). He received his AB degree in biochemistry, an MD degree, and a PhD degree in cell and developmental biology from Harvard University. He completed his clinical training in neurology at the University of California, San Francisco, and postdoctoral training in the Department of Molecular Pharmacology at Stanford University, where he was a Howard Hughes physician postdoctoral scholar. T.A.R. has received numerous awards and has been recognized by academic and professional societies for his work. He received a Paul Beeson Physician Faculty Scholar in Aging

Award from the American Federation for Aging Research and an Ellison Medical Foundation Senior Scholar Award. In 2005, he received an NIH Director's Pioneer Award for his work at the interface between stem cell biology and the biology of aging.

T.A.R.'s lab focuses on the biology of skeletal muscle stem cells in adult muscle homeostasis, aging and disease. In addition, his laboratory focuses on the pathogenesis and treatment of the muscular dystrophies, with particular emphasis on cell and gene therapy. His laboratory described the critical roles of Notch and Wnt signaling in the lineage progression of adult muscle stem cells, or satellite cells. Groundbreaking work from his lab showed that the age-related decline in stem cell function is due primarily to influences of the aged environment, rather than to intrinsic aging of stem cells themselves. These findings have profound implications for the emerging field of regenerative medicine, both for enhancement of endogenous tissue repair and for stem cell transplantation approaches. His laboratory continues to focus on adult muscle stem cells, with recent interests in epigenetic regulation and on the role of microRNAs in quiescence and activation.

Tannishtha Reya, PhD

T.R. is currently a tenured associate professor in the Department of Pharmacology and Cancer Biology and co-director of the Stem Cell Research Program at Duke University (www.reya-lab.org). She obtained her bachelor's degree from Williams College and received her PhD in immunology from the University of Pennsylvania. Her postdoctoral training began in Rudolph Grosschedl's lab at the University of California, San Francisco, where she demonstrated that lymphocyte development was dependent on LEF-1, a downstream activator of the Wnt signaling pathway. As a postdoctoral fellow in the laboratory of Irving Weissman at Stanford University, she made contributions to the field of stem cells and cancer. She continued this work in her own lab at the Department of Pharmacology and Cancer Biology at Duke University, where she was recruited in 2001. T.R.'s work to establish how diverse developmental mechanisms control both stem cells and cancer cells, creating opportunities to improve stem cell transplantation and cancer therapy, has been recognized by honors including the Presidential Early Career Award for Scientists and Engineers in 2005, the Leukemia and Lymphoma Scholar Award in 2007, the Endocrine Society's Richard E. Weitzman Memorial Award in 2008, and the NIH Pioneer Award in 2009.

The primary goal of the Reya lab is to understand the signaling pathways that regulate the choice between stem cell renewal and commitment, and to define how these mechanisms are dysregulated in cancer. The group is focusing in large part on developmental signals in stem cells and cancer, such as Wnt, Notch and Hedgehog, using molecular, cellular and transgenic approaches as well as real-time in vitro and in vivo imaging. T.R. was able to show that inhibition of developmental signals can block leukemia development and propagation in mouse models of the disease. The current work of the group is focused on further understanding the relationship between these pathways at a molecular level and also defining the role of these signals in human leukemias. Using models of regeneration based on delivery of chemotherapeutic agents and radiation, T.R. has begun to define the microenvironmental changes that occur after damage and which are sensed by stem cells to initiate the renewal process. Additionally, the Reya laboratory is investigating the intrinsic genetic program activated within stem cells that allows regeneration to occur, which will provide the basis for developing new approaches to accelerate regeneration after injury.

Hans-Reimer Rodewald, PhD

H.-R.R. carried out his PhD in immunology at the MPI for Immunobiology in Freiburg working on T cell receptor-biased T cell responses, and spent his postdoctoral years with Dr. Ellis Reinherz at the Dana-Farber Cancer Institute, Boston, focusing on T cell receptor isoforms and early T and NK cell development. In 1992, he started his own lab as a member of the Basel Institute for Immunology, where he identified T cell- and mast cell-committed progenitors, and characterized some growth factors (Kit) and cytokines that are crucial in lymphopoiesis, erythropoiesis, and myelopoiesis. Since 1999, H.-R.R. has been head of the Institute for Immunology at the University of Ulm (www.uni-ulm.de/klinik/immunologie).

The Rodewald group is focusing on thymus organogenesis, stem cell mutants, fate mapping of hematopoietic pathways, and mast cell mutants. H.-R.R. investigates the development and function of hematopoietic stem cells and the role of key growth factors and their receptors (c-kit) in stem cells by utilizing null mutants and rescue variants to study both physiological and pathological states of stem cells, and the role of c-kit in lymphopoiesis, erythropoiesis and myelopoiesis. The laboratory also studies the development and function of mast cells (allergy-related basic research) as well as thymus organogenesis. In the latter, an involvement of thymus epithelial 'stem' cells in medulla formation has been described. Moreover, techniques are available which allow mast cells to be generated from embryonic stem cells in vitro. Using this method, mutant mast cells can be generated to eventually analyze the role of certain gene products in allergic reactions.

Derrick J. Rossi, PhD

D.J.R. is an assistant professor in the Department of Pathology at Harvard Medical School. He has headed his lab at the Immune Disease Institute, in the Program of Molecular Medicine at Children's Hospital, Boston, since 2007 (www.hsci.harvard.edu). He is also a principal faculty member of the Harvard Stem Cell Institute. D.J.R. received his PhD from the University of Helsinki (Finland) in 2003, and did his postdoctoral work in the lab of Dr. Irving Weissman at Stanford University where he worked on hematopoietic stem cell biology and aging. D.J.R. was awarded a K99/R00 Pathways to Independence Award from the NIH and the National Institutes of Aging in 2006, and is also supported by grants from the Harvard Stem Cell Institute, the Helmsley Foundation, and GlaxoSmithKline.

D.J.R.'s research focuses on understanding the extent to which the aging of hematopoietic stem and progenitor cells contributes to the pathophysiological conditions arising in the aged hematopoietic system, which include declining immuno-competence, diminished regenerative potential, and cancer. To address this, the lab is studying the cellular and molecular mechanisms underlying the aging of the stem cells, with emphasis on the contribution of epigenetic regulatory mechanisms, in addition to mechanisms involved in the maintenance of genomic integrity in stem cells. The lab is also pursuing several lines of investigation aimed at reprogramming the cellular identity of a number of cell types into clinically useful cell types through various approaches including the use of novel technologies developed in the lab.

Karl Lenhard Rudolph, MD

K.L.R. heads the Institute for Molecular Medicine and the Max Planck Research Department on Stem Cell Aging at the University of Ulm (Germany) (www.uni-ulm.de/einrichtungen/instmolmed.html). After completing medical studies in Göttingen and a residency in internal medicine with Michael P. Manns in Hannover, he engaged in postdoctoral studies with Ronald

A. DePinho at the Albert Einstein College in New York, and at the Dana-Farber Cancer Institute in Boston. He started his own research group in Hannover in 2001, and moved to his current position in 2007. K.L.R. has received several research awards including the Gottfried-Wilhelm Leibniz Award of the DFG (the German Research Foundation) in 2009.

K.L.R.'s research focuses on the molecular mechanisms underlying the aging process, especially the aging of stem cells. He has carried out numerous projects on the functional consequences of telomere shortening in vivo. His main contributions include the demonstration that (1) telomere dysfunction can reduce organ maintenance, stress responses, and the lifespan of mice, (2) telomere dysfunction limits organ regeneration and is associated with the evolution of cirrhosis in humans with chronic liver disease, (3) telomere shortening leads to chromosomal instability and increased development of tumors, (4) the deletion of specific DNA damage checkpoints can improve stem cell function, organ maintenance, and the lifespan of telomere dysfunctional mice without increasing cancer formation, and (5) that telomere dysfunction induces environmental alterations limiting the function of stem cells.

Norman E. Sharpless, MD

N.E.S. is an associate professor of medicine and genetics at the University of North Carolina in Chapel Hill. N.E.S. went to medical school at the University of North Carolina, followed by a medical residency and oncology fellowship at Harvard Medical School's affiliated hospitals in Boston, Mass. (USA). After finishing his clinical training, N.E.S. completed a research postdoctoral fellowship in the lab of Dr. Ronald DePinho at the Dana-Farber Cancer Institute, and then joined the faculty at Harvard Medical School in 2000. N.E.S. returned to the University of North Carolina in 2002 where he runs a cancer genetics lab in the Lineberger Comprehensive Cancer Center (www.unclineberger.org). He was the 2007 recipient of the Jefferson Pilot Award, the 2009 recipient of the Hettleman Prize for Scholarly Achievement and was recently elected to the American Society of Clinical Investigation, the nation's oldest honor society for physician-scientists.

The Sharpless lab uses genetically engineered mice to study cancer and aging. His lab has shown that the p16INK4a tumor suppressor gene is both a biomarker and effector of aging in several mammalian tissues. Recent work from the lab has shown that measures of p16INK4a expression in circulating T cells can be used to estimate human molecular age.

Toshio Suda, MD

T.S. received his medical degree from Yokohama City University School of Medicine in Kanazawa-ku, Yokohama (Japan), followed by a residency at the Kanagawa Children's Medical Centre in Minami-ku, Yokohama, and a postdoctoral fellowship as a research associate at the Institute of Haematology, Jichi Medical School, Shimotsuke, Tochigi, and at the Department of Medicine, Medical University of South Carolina, Charleston. After returning to Japan, he worked as an assistant and then associate professor at the Jichi Medical School for 7 years, before being appointed a professor at the Institute of Molecular Embryology and Genetics' Department of Cell Differentiation, Kumamoto University School of Medicine. After 10 years in Kumamoto, he now works as a professor of developmental biology at the Sakaguchi Laboratory, School of Medicine, Keio University, Minato, Tokyo, Japan (www.coe-stemcell.keio.ac.jp).

T.S. was able to clarify the target cells for the colony-stimulating factors and to develop a new purification method for hematopoietic stem cells

'KSL cells' and progenitors. This is currently being used as a standard method for the purification of hematopoietic progenitors. He was also able to identify the osteoblastic niche for hematopoietic stem cells and subsequently established the new field of oxidative stress and stem cell aging. The interaction of stem cells and niche is considered to be one of the hot topics in stem cell biology. The findings of the Suda laboratory might be applied for stem cell transplantation and treatment of patients with bone marrow failure.

Andreas Trumpp, PhD

A.T. is full professor and head of the Department for Stem Cells and Cancer at the German Cancer Research Center (DKFZ) in Heidelberg, Germany (www.dkfz.de/en/stammzellen-und-krebs). He is also the founding managing director of the new Heidelberg Institute for Stem Cell Technology and Experimental Medicine (HI-STEM) GmbH. He studied biology at the University of Freiburg (Germany), receiving his diploma in 1989 and his PhD in 1992 at the European Molecular Biology Laboratories in Heidelberg (Germany). In 1994, he moved to the USA where he did postdoctoral research in the laboratories of J. Michael Bishop and Gail R. Martin at the University of California, San Francisco. He joined the Swiss Institute for Experimental Cancer Research (ISREC) in Epalinges/Lausanne as an associate scientist in 2000 where he became head of the Genetics and Stem Cell Laboratory. In 2005, he became professor (PATT) for molecular oncology and stem cell biology at the École Polytechnique Fédérale de Lausanne until moving to Heidelberg.

The Trumpp laboratory's goal is to use human patient samples and mouse molecular genetics to understand the molecular mechanism that controls the balance between stem cell self-renewal and differentiation, the molecular and cellular nature of the stem cell niche, the role of oncogenes and tumor suppressor genes in stem cells, the relationship between normal and cancer stem cells, and the role of disseminated tumor cells (metastasis stem cells) found in the circulation of cancer patients. The long-term goal of his work is to design novel strategies to detect and target cancer and metastatic stem cells.

Maarten van Lohuizen, PhD

M.v.L. received his PhD in 1992 from the University of Amsterdam, where he studied cooperating oncogenes in murine lymphomagenesis in the lab of Dr. A. Berns at the Netherlands Cancer Institute (NKI). After staying on for one year as a postdoc, he joined the group of Prof. Dr. Ira Herskowitz, University of California, San Francisco (USA) for his postdoctoral training. In 1995, he returned to the Netherlands Cancer Institute as an assistant professor in the Division of Molecular Carcinogenesis. After his tenure in 2000, he joined the Division of Molecular Genetics in 2001, of which he was appointed head of division in 2002 (http://research.nki.nl/lohuizenlab). In addition, in 2001 he became a part-time professor on the subject of regulation of cell cycle control and oncogenesis at the Utrecht University Medical School and was appointed as a member of the Centre for Biomedical Genetics in 2003. In 2007, he became part-time professor at the University of Amsterdam Medical School with the profile 'biology and epigenetic regulation of normal and cancer stem cells'.

His group has made important contributions on the functional analysis of epigenetic gene silencing mechanisms by Polycomb-group protein complexes, which play crucial roles in controlling development and cell fate and – when deregulated – contribute to cancer formation. Recently, his group has also developed high-throughput genome-wide genetic screens in cell-based assays and in cancer-prone mice (in collaboration with

Prof. Dr. A. Berns and Prof. Dr. A. Bradley of the Sanger Center, UK) to identify new genes that contribute to cancer and classify them in functional groups/signaling pathways. By using the DamID technique, his group has recently comprehensively mapped all Polycomb target genes in the *Drosophila* genome (in collaboration with the group of Bas van Steensel). In addition, his group has recently demonstrated a crucial role for Bmi1/Pc-G protein complexes in maintaining hematopoietic, neuronal, breast epithelial and embryonal stem cell fate and has implicated the Sonic Hedgehog (Shh) morphogen as a regulator of Bmi1 expression in neuronal precursor.

Unraveling the role of Bmi1/Pc-G in stem cell fate versus differentiation decisions and the consequence of this for cancer (stem) cell biology is currently a major focus of his group.

Gary Van Zant, PhD

G.V.Z. is a professor in the Departments of Internal Medicine and Physiology at the University of Kentucky School of Medicine. He holds the Lucille P. Markey Endowed Chair in Oncology Research at the Markey Cancer Center (http://ukhealthcare.uky.edu/Markey). He is director of the Clinical Stem Cell Laboratory for the Blood and Marrow Transplant Program of the Markey Cancer Center. G.V.Z. obtained his PhD degree from New York University and did his postdoctoral training at the University of Chicago. He held faculty positions at Washington University in St. Louis and the University of Michigan before joining the faculty of the University of Kentucky in 1995. G.V.Z. has received numerous research awards, including most recently a Senior Scholar Award from the Ellison Medical Foundation.

G.V.Z. has 30 years of experience studying hematopoiesis in general and stem cells in particular. He has been a leading proponent of the idea that the aging stem cell population in a variety of tissues and organs accounts, at least in part, for the decline in stress resistance of tissues, and perhaps to their longevity. G.V.Z.'s current research is focused on unraveling the regulatory pathways controlling maintenance and function of hematopoietic stem cells. In particular, his lab has used the natural diversity of inbred mouse strains to identify quantitative trait loci that affect stem cell aging, and alterations in the aging stem cell compartment that predispose to malignant transformation. For example, his lab recently identified latexin as a novel stem cell regulator, whose expression is age-dependent. Latexin is down-regulated in hematologic malignancies, and restoration of its expression in lymphoma cell lines diminishes their tumorigenicity in vitro and in vivo, thus demonstrating properties of a tumor suppressor. More recently, a similar genetic approach has identified a pocket protein of the retinoblastoma family, p107, as a regulator of hematopoietic stem cell aging.

Amy J. Wagers, PhD

A.J.W. directs a research laboratory at Joslin Diabetes Center (www.joslinresearch.org) and Harvard Stem Cell Institute (www.hsci.harvard.edu). She completed her PhD in immunology and microbial pathogenesis at Northwestern University in 1999 and received postdoctoral training in stem cell biology in Dr. Irving Weissman's lab in the Department of Pathology at Stanford University School of Medicine. She is currently an assistant professor in the Section on Developmental and Stem Cell Biology at Joslin Diabetes Center and in the Department of Stem Cell and Regenerative Biology at Harvard University. She is also a principal faculty member of the Harvard Stem Cell Institute and serves on its Executive Committee. A.J.W. is a recipient of the Burroughs Wellcome Fund Career Award, the Beckman Foundation Young Investigator Award, the WM Keck Foundation Distinguished Young Scholar

Award, and an Early Career Award from the Howard Hughes Medical Institute.

A.J.W. focuses on defining the factors and mechanisms regulating the migration, expansion, and regenerative potential of blood-forming and muscle-forming stem cells. Her lab employs sensitive cell sorting approaches for direct assessment of stem cell phenotype and function, and to develop surrogate in vitro assays that faithfully reflect stem cell potential.

Irving L. Weissman, MD

I.L.W. is a professor of pathology, developmental biology and, by courtesy, neurosurgery and biology at Stanford University. He is the Virginia and D.K. Ludwig professor for clinical investigation in cancer research and the director of the Stanford Institute of Stem Cell Biology and Regenerative Medicine, Comprehensive Cancer Center, and Stanford Ludwig Center for Stem Cell Research (http://med.stanford.edu, http://stemcell.stanford.edu and http://ludwigcenter.stanford.edu). I.L.W. started his scientific career at the McLaughlin Research Institute earning a BS from Montana State University in 1961 and received his medical degree from Stanford in 1965. After carrying out research in laboratories provided by Henry S. Kaplan, MD, he joined the faculty of Stanford 4 years later. His awards include the election to the National Academy of Sciences in 1989, the E. Donnall Thomas Prize to recognize pioneering research achievements in hematology in 1999, and being named California Scientist of the Year in 2002, the J. Allyn Taylor International Prize in Medicine in 2003, the New York Academy of Medicine Medal for Distinguished Contributions to Biomedical Research in 2004, the Jeffrey Modell 'Dare to Dream' Award in 2005, the Commonwealth Club of California 18th Annual Distinguished Citizen Award, the American-Italian Cancer Foundation Prize for Scientific Excellence in Medicine, the John Scott Award in 2006, the I. and H. Wachter Award in 2007 and the Robert Koch Award in 2008.

I.L.W.'s research encompasses the phylogeny and developmental biology of the cells that make up the blood-forming and immune systems. His laboratory identified and isolated the blood-forming stem cell from mice, and has defined, by lineage analysis, the stages of development between the stem cells and mature progeny. This required developing and cloning stromal cells of the hematolymphoid microenvironments – from the bone marrow for myeloid and B cells, and from the thymus for T cells. While the adhesion molecules and factors from these stromal cells proved important as molecules for myeloid and B cells, the analysis of T cell development required in vivo studies of thymic development. In addition, the Weissman laboratory has pioneered the study of the genes and proteins involved in cell adhesion events required for lymphocyte homing to lymphoid organs in vivo, either as a normal function or as events involved in malignant leukemic metastases. I.L.W. recently found a novel way to transplant new blood-forming stem cells into the bone marrow of mice without the tissue-damaging radiation or chemotherapy usually required, thereby effectively replacing their immune systems.

Ting Xie, PhD

T.X. is an investigator at the Stowers Institute for Medical Research (Kansas City, Mo., USA) and a professor of the Department of Anatomy and Cell Biology at the University of Kansas School of Medicine, Kansas City (www.stowers-institute.org). He obtained his PhD degree in molecular biology and biochemistry from Rutgers University under the supervision of Dr. Richard Padgett in 1996. He performed his postdoctoral research on *Drosophila* ovarian stem cell niche with Dr. Allan Spradling at Howard Hughes Medical Institute/Carnegie Institution of Washington during 1996–1999. T.X. started his own independent laboratory at the Stowers Institute for Medical Research in 2000 as an assistant investigator, and received the M.R. and Evelyn Hudson Foundation prize in 2003.

T.X. has been using *Drosophila* ovarian stem cells and, recently, mouse retinal stem cells in order to investigate the molecular mechanisms controlling niche function and formation, stem cell self-renewal, competition and aging. He and Dr. Spradling demonstrated, for the first time, that stem cell regulation is dependent on their niche. In addition, he has made important contributions in identifying key niche signals and providing a clearer understanding of how they collaborate with intrinsic factors to control self-renewal in the *Drosophila* ovary. His contributions include the demonstration that: (1) stem cells are kept in the niche through cadherin-mediated cell adhesion; (2) niche signals control self-renewal by repressing expression of differentiation-promoting genes; (3) the niche signal functions only within one cell diameter, allowing one daughter remaining in the niche to self-renew and the other daughter leaving the niche to differentiate; (4) ATP-dependent chromatin remodeling factors are required to work with niche signals to control self-renewal; (5) Notch signaling controls niche formation and maintenance, and (6) the microRNA pathway and translational regulators are required for self-renewal. Recently, he has successfully cultured a new population of stem cells from the adult mouse retina which can self-renew and generate, in vitro and in vivo, all the major retinal cell types and thus has potential in the treatment of retinal degenerative diseases, such as glaucoma.

Daniel Hartmann
Institute of Molecular Medicine
University of Ulm
DE–89081 Ulm (Germany)
Tel. +49 731 5036 112, Fax +49 731 5036 102
E-Mail daniel.hartmann@uni-ulm.de

Chapter 2

Rudolph KL (ed): Molecular Mechanisms of Adult Stem Cell Aging.
Else Kröner-Fresenius Symp. Basel, Karger, 2010, vol 1, pp 17–37

Hematopoietic Stem Cell Aging and Fate Decision

Anett Illing[a] • Zhenyu Ju[b]

[a]Institute of Molecular Medicine and Max Planck Research Department on Stem Cell Aging, University of Ulm, Ulm, Germany; [b]Stem Cell Biology and Aging Research, Sino-German Laboratory for Aging and Regenerative Medicine, Institute of Laboratory Animal Sciences, Chinese Academy of Medical Sciences, Beijing, PR China

Abstract
The following section will give an overview of the current data and theories on the aging of hematopoietic stem cells (HSCs). Aging is characterized by a gradual loss of regenerative capacity to sufficiently maintain tissue homeostasis and organ integrity. Mounting evidence suggests that a decline in functional stem cells can contribute to diminished homeostatic maintenance and regenerative potential in tissues during aging. Long-lived stem cells are prone to accumulate precancerous damage. Anti-tumor mechanisms restrain potentially malignant stem cell clones by inducing apoptosis or senescence, but simultaneously provoke aging. Clarification of molecular changes in aging HSCs will not only help our understanding of age-related bone marrow failure and immuno-senescence, but will also provide insights into the molecular basis for increased lympho-hematopoietic malignancies during aging. An understanding of these age-dependent processes could ultimately provide a rational basis for the development of new treatment and prevention strategies for age-associated bone marrow failure, immune dysfunction and leukemia development.

Characteristics of Hematopoietic Stem Cells

Hematopoietic stem cells (HSCs) are the best-characterized adult or somatic stem cells in the mammalian system [1]. HSCs ensure the maintenance and the development of all blood components, providing about 10^{11}–10^{12} new blood cells daily in humans [2]. A hierarchical structure in the hematopoietic system has been well documented. The most primitive stem cells reside at the apex of the hierarchy and are capable of both self-renewal and generation of progenitor cells that differentiate into a variety of mature cell types (fig. 1) [3, 4].

The most primitive HSCs are named long-term HSCs (LT-HSCs), which can be identified via a broad range of surface markers [5–11]. LT-HSCs are negative for lineage markers, negative for CD48, CD244, CD34, Flt3, and N-cadherin, but positive for Tie2, endoglin, Thy1 (preferably expressed at lower levels), Sca1, cKit, CD38 and CD150. Additionally, LT-HSCs, like several tumor cells, have been shown to express the P-glycoprotein, a member of the ABC-transporter family, enabling the stem cells to export substances like Hoechst 33342 shortly after uptake. Therefore, dye efflux ability can be used as an additional marker for LT-HSCs. In experimental setups, LT-HSCs are mainly sorted as $CD34^{low/-}$ $Flt3^{-}$KSL cells (low or negative for CD34, Flt3-negative, lineage-negative, Sca-1-positive and c-kit-positive). These cells are enriched for LT-

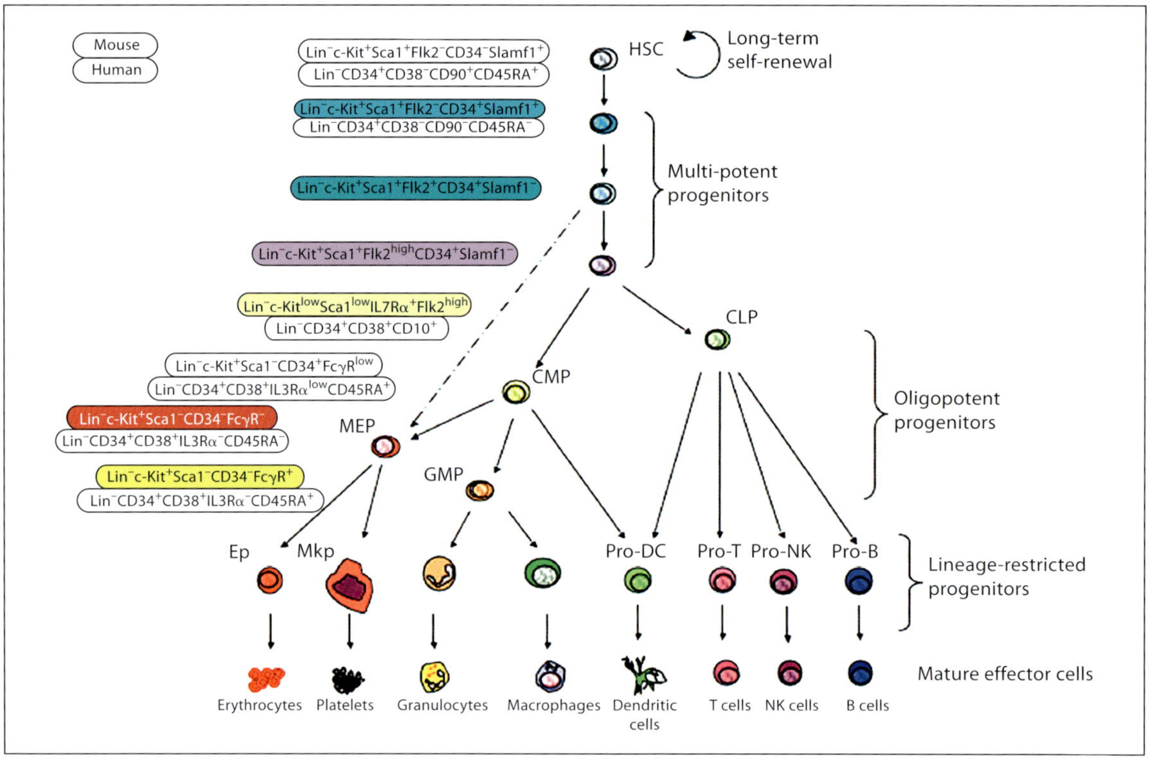

Fig. 1. Overview of the hematopoietic hierarchy and cell surface markers characterizing the different progenitor cells. CLP = Common lymphoid progenitor; CMP = common myeloid progenitor; GMP = granulocytic and monocytic progenitor; MEP = megakaryocytic and erythroid progenitor; Ep = erythroid progenitor; Mkp = megakaryocyte progenitor. Reproduced with permission from Weissman et al. [3].

HSCs (approx. 30–50% of the cells are LT-HSCs), having the potential to form all differentiated blood cell types via several progenitor stages and the ability of multi-lineage reconstitution in lethally irradiated mice [10]. In contrast to LT-HSCs, short-term HSCs (ST-HSCs) have already initiated differentiation and lack long-term reconstitution potential in serial transplantation experiments [12–16]. Multi-potent progenitor cells (MPPs) can give rise to all kind of progeny but lack self-renewal potential [9]. MPPs can differentiate into common myeloid and lymphoid progenitors [17, 18], which further differentiate to functional blood cells. It is difficult to distinguish MPPs from ST-HSCs. Gain of Flt3 expression has been proposed as a characteristic marker indicating the loss of self-renewal ability during HSC maturation [19]. Therefore, in addition to CD34, Flt3 represents another useful marker to prospectively isolate LT-HSCs, ST-HSCs and MPPs. Combined use of CD34, Flt3 and KSL markers gives rise to a classical phenotype pattern of long-term HSCs ($CD34^{low/-}Flt3^{-}KSL$), short-term HSCs ($CD34^{+}Flt3^{-}KSL$) and multi-potent progenitors ($CD34^{+}Flt3^{+}KSL$). Recently, SLAM (surface lymphocyte activation marker) markers have been discovered as powerful markers for identifying HSCs. By comparing changes in gene expression profiling in highly purified stem cells versus committed progenitor cells,

SLAM family members, including CD150, CD48, and CD244, were found to be differentially expressed at different stages of hematopoiesis in a way that associates with self-renewal potential [8]. At the Else Kröner-Fresenius Symposium on the Molecular Mechanisms of Adult Stem Cell Aging, independent lines of evidence came from 2 research groups, headed by Dr. Hiromitsu Nakauchi and Dr. Derrick Rossi, suggesting that CD150 is indeed a powerful marker for isolating most primitive LT-HSCs (see below), thus the most powerful, primitive HSC can be prospectively sorted by $CD150^{high}$ $CD34^{low/-}Flt3^{-}KSL$ phenotype.

Functional Changes during Aging

During the symposium, Prof. Gerald de Haan discussed current theories of HSC-aging. On the one hand, it is possible that HSCs age uniformly, with the complete stem cell population growing old. This is termed the 'population aging model' and for decades was thought to be true. Recently, opinions have changed and the theory of 'clonal selection' has been developed. According to this model, only a subset of HSCs undergo cellular aging. These altered cells accumulate over the lifetime, while the subpopulation of young HSCs gets constantly smaller. Different lines of evidence suggest that DNA damage induces checkpoints that may lead to a depletion of the most affected HSCs, although HSCs appear to be more resistant to damage accumulation than are progenitor cells (see chapter 9). It is also possible that different populations of HSCs show different rates of aging (possibly depending on their DNA damage or replication history). According to this model, different sub-populations of HSCs age at different speeds, leading to a selective depletion of some HSC subpopulations and an altered composition of stem cells during aging.

In this context, Christa Müller-Sieburg made an important contribution during the symposium. Her data support the hypothesis of clonal selection of HSC subpopulations during aging and indicate that there are 3 different populations of HSCs: (1) balanced HSCs (Bala); (2) lymphoid-biased HSCs (Ly-bi), and (3) myeloid-biased HSCs (My-bi). All types of HSC produce myeloid and lymphoid cells in blood; however, Ly-bi HSCs generate more lymphoid cells, and My-bi HSCs generate more myeloid cells than Bala HSCs. Surprisingly, these cells do not show aging phenotypes, as tested in limiting dilution transplantation experiments. However, the relative proportion of these individual subpopulations of HSCs changes during aging and, therefore, different reconstitution capacities of the total pool of HSCs occur. In conclusion, these data suggest that HSCs do not age in terms of losing their repopulation potential, but rather the My-bi HSCs accumulate with aging while the other 2 subpopulations of HSCs decrease, resulting in age-dependent skewing of hemato-lymphopoiesis potential with a decrease in lymphopoiesis and an increase in myelopoiesis. Furthermore, there is experimental evidence indicating that the My-bi HSCs from young mice are similar to those from aged mice regarding myelopoiesis, reduced lymphopoiesis, blunted IL-7 response, and repopulating capacity. The increase of myeloid-biased HSCs is a gene-driven phenotype. DBA2 mice have a shorter lifespan but higher numbers of HSCs at young age compared to C57BL/6J mice (see below). Although both C57Bl/6J and DBA/2 mouse strains show a reduction in Ly-bi HSCs and an increase in My-bi HSCs during aging, the short-lived DBA/2 mouse strain shows an earlier accumulation of My-bi HSCs compared to the C57BL/6J mouse strain. In addition, the increase in myelopoiesis was more severe in aged DBA2 mice (showing 80% My-bi HSCs) compared to aged C57Bl/6J mice (showing up to 60% My-bi HSCs). In summary, the selective survival of My-bi HSCs appears to be the main reason for impaired lymphocyte production in aging wild-type mice [20, 21].

This assumption was recapitulated by Dr. Nakauchi's and Dr. Rossi's experimental data indicating that $CD150^{high}CD34^{low/-}$KSL cells show myeloid dominant reconstitution, whereas $CD150^{low}CD34^{low/-}$KSL cells showed balanced or lymph-myeloid reconstitution. The $CD150^{high}$ $CD34^{low/-}$KSL cells have a molecular profile that appears myelo-erythroid primed, while $CD150^{low}$ $CD34^{low/-}$KSL cells are more lymphoid primed. The greater self-renewal potential of myeloid-biased $CD150^{high}CD34^{low/-}$KSL cells resulted in their expansion in the primitive stem cell pool during aging. These data support the model of clonal selection of subpopulations of HSCs during aging; however, we cannot conclude that the increase of My-bi HSCs in aged mice is solely due to population shift. In fact, most probably both clonal selection and population shift take place in aged mice. Further study is necessary to make a clear conclusion on this issue.

Although serial transplantation experiments were often used as a standard approach to evaluate stem cell self-renewal, Dr. Irving L. Weissman mentioned that such experiments could artificially be complicated by the effects of irradiation. Dr. Weissman explained that irradiation might create an environment in the host bone marrow that promotes the cell cycle of engrafted HSCs and probably favors the selection of My-bi HSCs. It is known that cycling HSCs do not engraft well, therefore a stochastical loss of cycling HSCs during serial transplantation could occur. In contrast, in $Rag^{-/-}$ mice pre-conditioned with ACK2 (anti c-kit targeting monoclonal antibody), which wipes out the HSCs without irradiation damage, HSCs immediately go back to a low cycling status once they have reconstituted the stem cell pool [22].

The molecular mechanisms that lead to stem cell aging or the selection of subpopulations of stem cells during aging remain to be defined. One possibility is that there is no perfect self-renewal in adult HSCs or in certain subpopulations of HSCs. Taking advantage of the known HSC markers (see above), HSCs were prospectively isolated and purified from C57BL/6J mice for evaluating the impact of aging on the stem cell compartment using single-cell repopulation experiments and limiting dilution assay [23, 24]. These studies suggested that aging was accompanied by a steady increase in the frequencies of primitive hematopoietic cells, but the repopulating capacity of aged HSCs was reduced on a per cell basis. These observations elicit a model of stem aging – clonal selection – that certain stem clones bearing distinct functional potentials are able to out-compete other types of clones during aging (fig. 2). Over time, such clones will predominate the stem cell pool, whereas others become diminished.

Immuno-Phenotype Changes during Aging

Data on serial transplantation indicate that the replication history restricts the function and lifespan of HSCs. However, it is still not clear whether replicative stress has an impact on the immuno-phenotype or the renewal capacity of HSCs during aging.

During the symposium, Dr. Nakauchi suggested that division history could affect both the functionality and the immuno-phenotype of HSCs. One way to dissect this question would be the analysis of single, phenotypically defined HSCs. The discovery of SLAM markers has fueled this argument. By comparing changes in gene expression profiling in highly purified stem cells versus committed progenitor cells, SLAM family members including CD150, CD48 and CD244 were found to be differentially expressed at different stages of the hematopoiesis hierarchy in a way that associates with self-renewal potential [8]. It was originally discovered in young adult bone marrow that CD150 is expressed by HSCs, while CD244 is expressed by non-self-renewing MPPs, and CD48 is expressed by restricted progenitors. The SLAM family receptors were conserved among HSCs from diverse contexts,

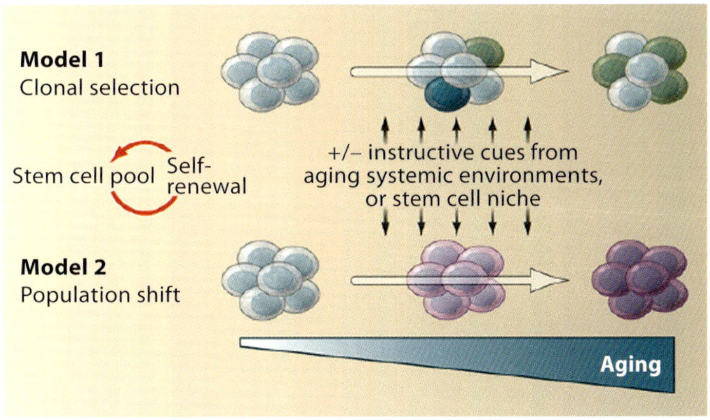

Fig. 2. Current models of stem cell aging [25]. Model 1 – clonal selection – suggests that certain stem clones bearing distinct functional potentials are able to out-compete other types of clones during aging. Over time, such clones will come to predominate the stem cell pool, whereas others become diminished. Model 2 – population shift – suggests that the functional potential of all of the cells in the stem cell population is equivalent at any stage of ontogeny, but this potential gradually and coordinately changes over time. These models are not mutually exclusive and both may be instructed by non-cell autonomous cues originating from the aging micro- or systemic environment. Reproduced with permission from Rossi et al. [25].

including old, reconstituted, or mobilized mice, and markedly increased the maximal purification of HSCs [26]. Dr. Sean J. Morrison thinks that including the CD48 marker could clean out non-functional HSCs that appear to accumulate in the CD34$^{low/-}$KSL gate during aging of C57Bl/6J mice. To address this question, Dr. Rossi and Dr. Nakauchi went further to dissect the function of CD34$^{low/-}$Flt3^{-}KSL cells according to SLAM marker expression. Among their findings, Dr. Rossi provided data showing that CD34$^{low/-}$Flt3^{-}KSL HSCs are CD48^{-} regardless of the mice age. These data argue against the hypothesis that CD48^{+} progenitors accumulate in the CD34$^{low/-}$KSL gate during aging, as has been shown for less purified KTLS (c-kit^{+}Thy-1^{low}Lineage^{-}Sca-1^{+}) subset during aging [26].

Dr. Yohei and Dr. Ema from Dr. Nakauchi's lab have been devoted to testing many new markers to enhance the purification efficiency for HSCs. Competitive transplantation experiments revealed that some of the tested markers, such as Tie2, CD41, CD150, CD38 and CD147 showed some difference in repopulating capacity after transplanting 10 of these cells into irradiated mice. Among them, CD150 was the most powerful marker to isolate the most primitive HSCs, as tested by single cell transplantation and in vitro colony formation. When dividing CD34$^{low/-}$KSL cells into negative, low and high compartments according to CD150 expression, the CD150^{high}, CD150^{low} or CD150^{-} cells in the CD34$^{low/-}$KSL cell compartment showed similar engraftment rates, but the engrafting kinetics were different. The peripheral blood chimerism of CD150^{low} CD34$^{low/-}$KSL donor-derived cells was the highest in transplanted recipients. With time, the chimerism of CD150^{-}CD34$^{low/-}$KSL donor-derived cells declined, whereas that of CD150^{high}CD34$^{low/-}$KSL donor-derived cells showed a gradual increase. One tenth of the CD150^{high}CD34$^{low/-}$KSL cells showed very slow engraftment; the peripheral blood chimerism showed up at 4–5 months after transplantation, which is consistent with Prof. Andreas Trumpp's finding that the most primitive HSCs have a very slow cell cycle. In the second transplantation of 5 million bone marrow cells from the first recipients, CD150^{high}CD34$^{low/-}$KSL donor-derived cells showed increased engraftment, while CD150^{low}CD34$^{low/-}$KSL showed very poor engraftment and CD150^{-}CD34$^{low/-}$KSL donor-derived cells showed almost no engraftment in second recipients. Moreover, the CD150 phenotype appeared to be hierarchical as

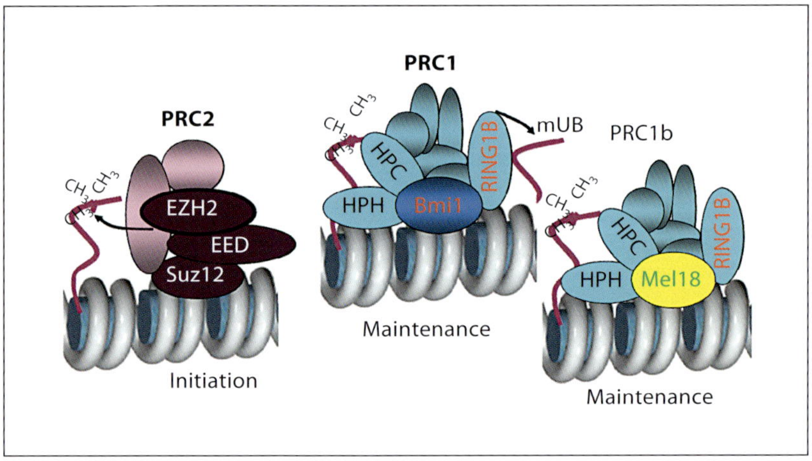

Fig. 3. Composition of the PRC1 and PRC2 complex in the context of the initiation or maintenance of methylation marks. The composition of the PRC2 complex, including EZH2, EED and Suz12, is essential for the initiation of methylation marks on the DNA. Alternative compositions form the PRC1 or PRC1b complex. Both protein complexes, PRC1 and PRC1b, are responsible for maintaining methylation marks on the DNA, for example during replication of DNA. PRC1 and PRC1b differ only in one component: PRC1 includes Bmi1, whereas PRC1b includes Mel18. Bmi1 and Mel18 are exchangeable in vitro but have diverse functions in the regulation of adult stem cells. Scheme: reproduced with permission from M. van Lohuizen.

$CD150^{high}CD34^{low/-}$KSL could generate $CD150^{low}$ $CD34^{low/-}$KSL cells, whereas $CD150^{low}CD34^{low/-}$ KSL cells could not generate $CD150^{high}CD34^{low/-}$ KSL cells. These findings were independently supported by the Rossi group.

Epigenetic Regulation of Stem Cell Function

Recently, the hypothesis of an epigenetic deregulation of gene expression in aging stem cells has emerged [27, 28]. According to this hypothesis, the loss of epigenetic control might contribute to stem cell aging. In detail, the loss or gain of organization in hetero- or euchromatin, or relocalization of heterochromatin would result in differentially expressed transcripts or completely random expression. Likely candidates for epigenetic regulation are members of the polycomb (PcG) family. PcGs were first identified as a conserved group of genes in *Drosophila* that are required in multimeric protein complexes to maintain the expression of *Hox* genes [29–32]. PcG proteins interact with chromosomal elements, termed cellular memory modules. They have a repressive function, which is stable over generations and can only be overcome by germline differentiation processes.

Several mammalian homologues of *Drosophila* PcGs have been identified. PcG proteins have been shown to form 2 complexes, PRC1 and PRC2 (fig. 3). Both complexes are necessary for the repressive effect. PRC2 initiates PcG-mediated repression, whereas PRC1 maintains the repression.

PRC2 includes the histone-methyltransferase EZH2 setting trimethyl marks on lysine 27 of the histone H3-tail. This repressive mark is recognized by the chromodomain proteins in the PRC1 complex [33, 34]. PRC1 is larger than PRC2 and has enzymatic activity as an E3-ubiquitine ligase. This is mediated by RING1B, which is forming a heterodimer with Bmi1 acting as a regulator and enhancer of the complex.

The mono-ubiquitine mark is set on histone H2A. This mark might be important for PcG repression and it keeps the polymerase II stalled at the PcG targets. All chromatin marks are highly dynamic and can be removed by demethylases. Depending on the context, the marks are continuously set or removed [32, 35–43].

Bmi1 and Mel18 in the PRC1 complex were shown to be exchangeable. Because of their 75% similarity, they display exactly the same biochemical functions in vitro. The knockout phenotypes overlap but they are not identical; for example, Bmi1 and Mel18 display distinct functions in the regulation of adult stem cells. Bmi1 is essential in the maintenance and the regulation of self-renewal in HSCs. $Bmi1^{-/-}$ mice have markedly reduced numbers of HSCs which are defective in self-renewal and can only transiently contribute to hematopoiesis [47]. Just recently it was proven that Bmi1 alone is not the only regulator of self-renewal but a well-balanced expression of Bmi1 and Mel18 is required to maintain stem cell differentiation and self-renewal [48].

It is known that there is a profound failure of Bmi1-deficient HSCs to reconstitute irradiated recipients in contrast to wild-type HSCs [47]. Moreover, Bmi1 is not only required for the self-renewal and reconstituting activities of normal HSCs but is also required by leukemic stem cells [49]. Bmi1-deficient hematopoietic cells transformed by Hoxa9 and Meis1 (a homeobox protein, frequent co-expression of Hoxa9 and Meis1 genes was found in infant acute lymphoblastic leukemia with MLL rearrangement) could form acute myeloid leukemia. However, the leukemic cells of $Bmi1^{-/-}$ HSCs were not transplantable, whereas Hoxa9/Meis1-induced leukemia from $Bmi1^{+/+}$ HSCs was transplantable [49].

During the symposium, Dr. Morrison showed new data on identifying genes that limit self-renewal of Bmi1-deficient HSCs. He used forward genetic somatic screening by sleeping beauty transposon mutagenesis in hematopoietic cells to look for Bmi1 suppressors. The screen identified the Prdm16 gene to be required for the maintenance and function of stem cells. The expression pattern of Prdm16 was highly restricted to stem cells in the hematopoietic system.

Dr. Maarten van Lohuizen also discussed the role of Bmi1 in stem cells, cancer and development. Bmi1 over-expression cooperates with loss of phosphate-tensin homolog (PTEN) in promoting prostate carcinogenesis by depressing the Ink4a/Arf tumor suppressor pathway (see chapter 9). Regarding the role of Bmi1 in stem cells, studies on Bmi1-deficient mice showed overall growth retardation, premature death, posterior transformation, neurological abnormalities and severe hematopoietic defects [50]. Further investigation revealed that Bmi1 is required for the self-renewal and maintenance of both HSCs and neural stem cells [47, 51]. Dr. van Lohuizen found that Bmi1 controls differentiation in mammary epithelial stem cells and progenitor cells. Mammary tissue from Bmi1-deficient mice displayed a premature differentiation leading to a dramatic loss of repopulating capacity of mammary stem cells [52]. Loss of $p16^{Ink4a}/p19^{Arf}$ (also known as Cdkn2a) rescued the phenotype in mammary tissue transplantation experiments. Together with evidence from other stem cell compartments, the de-repression of the $p16^{Ink4a}/p19^{Arf}$ locus was suggested to be responsible, at least in part, for the abnormalities developed in Bmi1-deficient mice [53–55]. Although the deletion of $p16^{Ink4a}/p19^{Arf}$ in Bmi1-deficient mice improved stem cell function, the cellular composition of the bone marrow microenvironment remained significantly altered in double-null mice [56]. In addition, the body weight and overall survival remained reduced compared to wild-type littermates. In the symposium, Dr. Toren Finkel reported a new pathway involving reactive oxygen species production that is responsible for the abnormalities in Bmi1-deficient mice (see chapter 8).

The above-mentioned methylation of lysine 27 residue on histone H3 (Lys27H3) was shown to be important in HSCs. This led to the hypothesis that PcGs are able to regulate the heterogeneity of HSCs [46]. Methylation of Lys27 on H3 leads to suppression of secondary genes. The methylation mark can be copied during cell division and the repressed state is inherited by the daughter cells. Prof. de Haan hypothesized during the symposium that PcGs could be absent or less abundant in HSCs. This could lead to a less faithful copying of methylation marks on Lys27H3 in dividing stem cells compared to somatic cells. The proposed mechanism would increase the variability of HSCs in a 'not-programmed fashion'. However, this hypothesis needs to be tested. In addition, it is possible that stem cell-specific characteristics of DNA replication (e.g. maintenance of an immortal strand) would impact on the maintenance of methylation marks. It is also possible that aging itself would affect the fidelity of methylation mark copying in dividing stem cells. Along these lines, the PRC1 complex was shown to display differences in its activity to copy methylation marks in young versus old cells, involving an altered recruitment of the PRC1 complex to silenced loci. This process resulted in an increased heterogeneity of cell populations regarding the methylation state [57].

In this context, it is important to also mention the presentation of Dr. Hans Jörg Fehling at the symposium. The counterplayers of PcGs are Thritorax group proteins (Trx-G) that were shown to maintain the expression of Hox and other developmental genes by epigenetic modification of histones [58]. Trx-G and PcGs control the maintenance, but not the initiation, of gene expression as epigenetic guardians of cell identity. Furthermore, there is also evidence for the involvement of Trx-G in cellular proliferation and tumorigenesis [59–61]. Fehling and colleagues found that Mll5, the latest MLL/Trx-G member, impairs the function of primitive hematopoietic cells. Mll5 is expressed in most somatic tissues. To obtain a first understanding of potential physiological functions, Mll5-deficient mice were generated using gene targeting in embryonic stem cells and Cre/loxP technology. Intercrosses of heterozygous Mll5$^{+/-}$ mice produced viable homozygous KO/KO offspring, however at significantly reduced numbers. Further studies revealed that approximately 40% of homozygous Mll5-deficient pups died for unknown reasons during the first days of postnatal life. Mll5$^{-/-}$ pups surviving this critical period developed into outwardly healthy adults. However, further analyses revealed a multitude of phenotypic abnormalities, including complete male sterility, impaired female fertility, retarded growth and defective lymphopoiesis. For instance, B and T cell numbers were greatly reduced, although there were no defects in T cell receptor recombination. Double-positive and double-negative cells in the thymus were reduced proportionally, while CD117-positive early thymocyte progenitors were almost absent. Additionally, the total number of LSK cells, which contain hematopoietic stem/progenitor cell populations, was decreased in Mll5 knockout mice and lineage-negative cells showed strong reconstitution defects in transplantation experiments. The fact that Mll5$^{-/-}$ mice exhibited hematopoietic failure after ionizing irradiation (sublethal dose) led to the conclusion that Mll5 plays a vital role in the response to damaging γ radiation, which is, however, restricted to specific cell types, including the hematopoietic stem and progenitor populations [62].

DNA Damage Signaling in the Regulation of HSCs

One of the mechanisms of aging is the accumulation of DNA damage [63]. Activation of the DNA damage response correlates with impairment of stem cell function. Studies of mouse strains bearing mutations in DNA damage response or repair pathways strengthen the concept that DNA damage can impair the functional reserve of stem cell compartments [64–67]. Using comet assay and γ-

H2ax staining, Dr. Rossi found that DNA damage accumulates in aged HSCs and is restricted to the stem cell compartment. Dr. Rossi suggested that this could be due to an attenuated DNA repair activity in HSCs compared to downstream progenitors. Indeed, global expression profiling revealed that the DNA repair and response pathways are down-regulated in quiescent but not in cycling HSCs. HSCs cycle slowly and show little DNA repair activity, but damaged stem cells can repair DNA damage when they re-enter the cell cycle [67]. Dr. Rossi provided experimental evidence that HSCs that harbor unrepaired DNA damage can also be cleared out by apoptosis and possibly, in rare circumstances, by senescence. One molecular mechanism that could be involved in the induction of stem cell senescence is the epigenetic de-repression of the Ink4a/Arf locus (see above), although recent evidence suggests that this mechanism may not be active in HSCs. Deletion of $p16^{Ink4a}$ leads to improved functional capacity of neural, pancreatic and hematopoietic stem cells [68–70]. In addition to DNA damage and de-repression of $p16^{INK4a}$, telomere shortening represents another mechanism mediating cellular senescence, growth arrest and/or cell death during aging [71, 72].

Telomere dysfunction represents a specific form of DNA damage that cannot easily be repaired. In the absence of functional telomerase, telomere attrition both attenuates stem cell function and contributes to organismal aging in mice or humans. Late-generation telomerase-deficient mice have a reduced lifespan and experience progressive degeneration in proliferative tissues such as skin, intestine and blood [73–75]. The impact of telomere dysfunction on stem cell compartments in telomere-dysfunctional mice depended on both cell intrinsic checkpoints and cell extrinsic alterations [76–79]. The molecular mechanisms of stem cell defects in response to telomere dysfunction will be discussed in other chapters of this book (see chapter 9). Briefly, when telomeres become critically short, tumor suppressor pathways mediated by p53 are activated, leading to senescence or apoptosis. The CDK inhibitor p21 is one of the downstream effectors of p53 that is responsible for cell cycle arrest. Different lines of evidence indicate that the deletion of p21 improves the maintenance and function of telomere-dysfunctional stem cells, and extends the lifespan of late-generation telomerase-deficient mice [76, 80]. Dr. Karl Lenhard Rudolph reported a novel concept indicating that p53-dependent checkpoints have a protective role for tissue homeostasis in the context of telomere dysfunction by mediating clearance of chromosomally unstable stem cells with critically short telomeres [81].

During the symposium, Dr. Tao Cheng presented new experimental data showing that inactivation of Puma protects stem cells in response to irradiation, resulting in improved long-term survival of irradiated $Puma^{-/-}$ mice compared to $Puma^{+/+}$ mice. Puma was originally identified as a direct p53 target gene. It encodes a BH-3-only cell death protein and appears to be essential for cell death triggered by ionizing radiation, deregulated c-Myc expression and cytokine withdrawal. It has been reported that lymphoid cells and myeloid progenitors are resistant to irradiation in the absence of Puma [82]. A recent study documented a significant role for Puma for the induction of apoptosis in response to DNA damage in mouse intestinal progenitor cells [83]. The baseline expression of Puma in LT-HSCs was relatively low compared to other subsets of hematopoietic progenitor cells (ST-HSCs, MPPs or committed progenitors). Dr. Cheng provided new data on the hematopoietic system that showed that the expression of Puma in LT-HSCs increased threefold within 2 h after 8-Gy irradiation. Deletion of Puma did not alter HSC homeostasis in non-irradiated mice, but protected KSL cells in response to γ irradiation. In competitive transplantation experiments, 4-Gy irradiation resulted in a strong selective advantage of $Puma^{-/-}$ HSCs compared to $Puma^{+/+}$ HSCs in both primary and secondary recipients. In addition, recipient mice transplant-

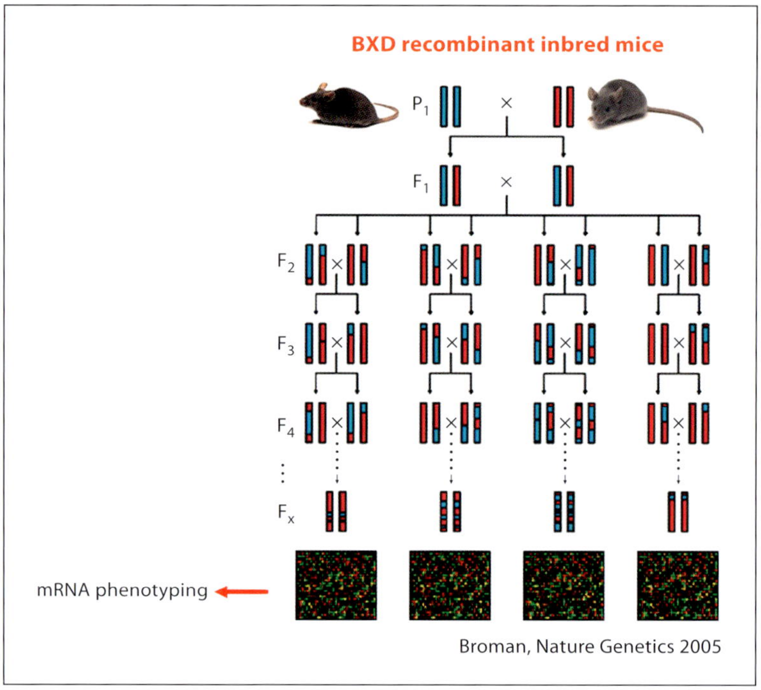

Fig. 4. Overview of the creation of BXD mouse strains, taken from Broman [88]. BXD mice are created by crossing 2 different mouse strains that differ in one or more phenotypic attributes. By inbreeding the offspring for several generations, mice with a mosaic genome are created. The last generation of these mice differs only in small genomic regions which might be linked to a certain phenotype. Reproduced with permission from Broman [88].

ed with Puma$^{-/-}$ HSCs showed a strong radioresistance compared to recipients transplanted with Puma$^{+/+}$ HSCs. After 2 rounds of irradiation, the recipient mice reconstituted with Puma-deficient HSCs survived over 100 days, whereas the controls died within 20 days. In the absence of Puma, up-regulation of p21 in response to irradiation preserved the stem cell quiescence. In addition, Dr. Cheng presented data indicating that Puma deletion can enhance the DNA damage repair in response to irradiation compared to Puma$^{+/+}$ mice.

Dissecting Age-Dependent Effects on Hematopoiesis in Recombinant Inbred Mouse Strains

One interesting genetic tool to investigate molecular components of HSC aging are recombinant inbred mouse strains. The BXD strains are recombinant inbred mice that are obtained by crossing 2 different mouse strains together (fig. 4) [84]. At the symposium, data were presented that had been obtained from mice generated by crossing C57Bl/6J mice with DBA/2 mice, followed by repeated inbreeding of the offspring for several generations. The genome of the resulting new inbred line is a mosaic of the 2 parental strains. With this method, phenotypic differences of 2 strains can be investigated and linked to specific chromosomal regions that are influencing the phenotype. This so-called quantitative trait loci (QTLs) analysis, identifying the part of the chromosome that influences the phenotypic property, was initially used in evolutionary biology and to identify disease-responsible genes in humans. The identification of QTLs is based on variations of genetic markers like restriction fragment length polymorphisms, single nucleotide polymorphisms, short tandem repeats or copy number variations, which can vary in different individuals or between populations. QTLs are also used to identify the basis of gene expression variations due to the fact that QTLs affect transcript abundance. In this coherence, *cis*- and *trans*-acting effects have to be discriminated. If transcripts are influenced by a locus at or near its own genetic location, this is named a *cis*-acting QTL, whereas other transcripts are affected by QTLs that are in completely different genomic locations than the transcript itself (*trans*-acting QTLs) [85–87].

Prof. Gary Van Zant and coworkers have used BXD recombinant inbred strains to identify factors influencing HSC aging. The numbers of HSCs in young B6 mice are known to be relatively low as compared to D2 HSCs [89, 90]. Furthermore, B6 cells show lower proliferation and mobilization but, surprisingly, have a higher self-renewal capacity, and the number of HSCs increases during aging and does not decrease as seen in aging D2 mice. To investigate the variation between the 2 strains, embryo-aggregated chimeric mice were created. These cellular chimeras were analyzed in aging studies during their entire lifetime for hematopoietic phenotypes. There was a strong selection of B6 stem cells and the percentage of D2 HSCs contributing to hematopoiesis decreased during aging of the chimeric mice [91]. However, the D2 HSCs did not die out but became quiescent and could be reactivated by bone marrow transplantation. It was concluded from these data that HSCs of B6 and D2 mice have clearly different aging patterns. The D2 animals show the highest numbers of HSCs at about 1 year of age, after which stem cell numbers decline. B6 mice display a constantly increasing HSC pool throughout life. Although the actual number of HSCs increases with aging in the B6 strain, the regenerative potential of individual HSCs decreases.

To identify responsible chromosomal regions or even genes for these phenotypic differences in HSC number during aging, a forward genetic approach was initiated, beginning with a genome-wide linkage analysis using a panel of BXD recombinant inbred mouse strains. A QTL for the phenotype was mapped to distal chromosome 2 [unpubl. data]. For further investigations, congenic mice were generated in which a segment of chromosome 2 containing the QTL was moved by backcrossing onto the B6 genetic background. Phenotypic analysis of the congenic strain showed that the numbers of HSCs in the young mice did not differ from pure B6 controls. However, 24-month-old congenic mice had only half of the number of HSCs compared to pure B6 not containing the D2 locus on chromosome 2. This result reconfirmed that the D2 allele on chromosome 2 was in fact a genetic modifier controlling HSC self-renewal in aging mice. It was subsequently found that the D2 allele also increased radio-sensitivity of HSCs.

A current aim in the laboratory of Prof. Van Zant is to find the phenotype-specific gene(s) in the chromosome 2 QTL that influence the self-renewal of HSCs during aging. Expression profiling of HSCs and MPPs for differentially expressed genes in congenic and B6 mice revealed p107 as one such gene. p107 is located within the locus on chromosome 2. Interestingly, previous studies have shown that p107 can regulate the maintenance of neural stem cells [92]. p107 belongs to the pocket protein family and is a retinoblastoma-like protein acting as a tumor suppressor. It is involved in the blockage of E2F-responsive genes that are involved in cell cycle progression. The expression of p107 is *cis* regulated, so its promoter accounts for most, if not all, expression of p107 [93–95]. Prof. Van Zant and colleagues found that the B6 allele is associated with an increased expression of p107 compared to the D2 allele. The regulation of HSC numbers by p107 was confirmed in knockdown and over-expression experiments. The promoter region and the 3′ end of the p107 gene show evidence for a regulation of the gene by microRNAs. Comparisons of the B6 and D2 alleles show that these regions display several single nucleotide polymorphisms. It is possible that these polymorphisms influence the expression of p107. In addition, FOXO transcription factors could be involved, as it is known that the FOXO proteins regulate p107 expression and cell cycle activity [96].

Prof. Van Zant and coworkers also used the BXD strains to identify QTLs that are related to HSC phenotypes. Mice were generated by crossing the genomic interval harboring the QTL using D2 as the donor strain and C57Bl/B6 as the recipient strain. The 2 congenic strains were derived by a 'speed-congenic' approach involving at least 8 backcrosses with the respective background strains [97, 98]. During the symposium, Prof. de Haan presented data on the identification of a region on chromosome 3 that was significantly associated with HSC aging, namely with altered HSC numbers. This region on chromosome 3 includes the latexin gene. Latexin is highly expressed in the most primitive population of HSCs and is down-regulated during differentiation. Independent of the type of hematopoietic cells, latexin levels are always higher in cells from C57Bl/6J mice compared to D2 mice. The B6 allele of latexin results in lower HSC numbers, shown by numerical and functional analysis, compared to D2 mice. Several experiments confirmed that latexin regulates the self-renewal capacity of HSCs and that this is a cell-intrinsic effect. Latexin is *cis* regulated because its regulatory region, D3Mit5, is closely upstream of the gene. Like the intron regions of the gene, the 5′-and 3′-untranslated regions were shown to contain several single nucleotide polymorphisms, possibly explaining the inter-strain variations in gene expression [99].

Prof. de Haan presented new data indicating that mouse strain-specific modifiers can affect the function of genes regulating stem cell biology. He showed his group's findings on the cyclin-dependent kinase inhibitor p21. Deletion of p21 in mice of the 129/SvEv background results in an increase in HSC cycling and in the total number of HSCs in young mice. However, p21 deletion in 129/SvEv mice is associated with premature stem cell exhaustion in serial transplantation experiments and with an increased chemosensitivity to 5-fluorouracil (5-FU) treatment [100]. In contrast to these effects of p21 on stem cell quiescence in 129/SvEv mice, Prof. de Haan showed that a knockout of p21 in the C57Bl/6J background shows only minimal effects [101] (for details see section below). In the end, it will be important to analyze whether the findings from knockout experiments in specific inbred mouse strains translated to human biology. The findings on p21 are even more complex when we look at the role of p21 in the context of telomere dysfunction-induced aging (chapter 9). It was shown that p21 deletion had no significant effects on stem cell aging in wild-type C57Bl/6J mice. In contrast, the deletion of p21 improved stem cell maintenance and function in telomere-dysfunctional ($Terc^{-/-}$) C57Bl/6J mice. These data indicated that p21 not only has positive effects on the maintenance of stem cell function (as observed in D2 mice), but can also mediate the negative effects on stem cell self-renewal and function induced by DNA damage signaling in response to telomere dysfunction [76].

A different example for mouse strain-specific changes in the expression of HSC-regulating genes is the Slit2 gene. The mammalian SLIT genes belong to a highly conserved family of axon guidance molecules. Recently, they were also shown to play a role in acute and chronic lymphoid leukemias [102, 103]. Prof. de Haan and colleagues showed that Slit2 is cell type-specific and the most primitive population of HSCs expresses high levels of Slit2, whereas committed progenitor cells express significantly lower levels of the gene. Slit2 expression also shows differences in D2 compared to B6 mice. The latter mouse strain possesses low expression levels of Slit2 compared to C57Bl/6J mice [104]. Moreover, Prof. de Haan showed that 182 genes were strain-dependent regulated in HSCs of D2 compared to

B6 mice. The question arises whether there are specific transcription factors that regulate these genes. In this context, it could be important that gene expression in differentiated cells is quite often *trans* regulated, whereas gene expression in stem cells (including HSCs) is rather *cis* regulated [105, 106]. In addition, there remains a critical question: which of the strain-specific differences in gene expression – which influence HSC function, self-renewal and aging – is important for human aging, specifically in regard to stem cell aging?

Genes Controlling HSC Quiescence and Self-Renewal

Studies on a wide range of genetically manipulated model systems demonstrated that intrinsic HSC alterations can lead to the dysfunction of stem cells. PTEN, a negative regulator of the PI3K-Akt pathway and tumor suppressor, is crucial for the inhibition of leukemogenesis. If PTEN is lost, ST-HSCs are strongly activated and LT-HSCs are depleted over time. PTEN-deficient HSCs engraft with normal efficiency after transplantation but hematopoietic reconstitution cannot be sustained due to a deregulation of the cell cycle. Loss of PTEN leads to myeloproliferative disease and transplantable leukemia, since PTEN functions to restrict the proliferation of hematopoietic progenitor cells [107, 108].

Related to the function of PTEN is the cellular oncogene c-Myc. Loss of c-Myc leads to accumulation of LT-HSCs in the niche due to up-regulation of adhesion molecules, especially N-cadherin. It has been shown that this increased adhesion impairs the exit of LT-HSCs from the niche. Conversely, over-expression of c-Myc leads to the loss of HSCs due to premature differentiation associated with a down-regulation of adhesion molecules. In summary, c-Myc is an important player in regulating the fate decision between self-renewal and differentiation [109].

As discussed above, cyclin-dependent kinase inhibitors also have a function in the regulation of stem cell quiescence and stem cell maintenance. The individual knockouts for p18 and p21 show a strong HSC phenotype, both resulting in increased proliferation of HSCs. The deletion of p18 results in an increase of the functional HSC pool [110], whereas p21 deletion can lead to an early exhaustion of LT-HSCs in certain genetic backgrounds (see above). Together, it appears that the maintenance of stem cell quiescence can contribute to stem cell maintenance. However, the differential effects of p18 and p21 on stem cell maintenance indicate that factors other than maintenance of quiescence could be involved. Another interesting observation was that p18 deletion rescued the effects of p21 deletion on HSC depletion [111].

Two other important intrinsic HSC regulators are myeloid elf-1-like factor (MEF/ELF4) and early growth response 1 (Egr1). MEF/ELF4 regulates the quiescence of LT-HSCs. Deletion of MEF leads to enhanced quiescence of HSCs. The reconstitution potential of MEF-deleted HSCs is normal and the mice show an enhanced protection against myelotoxic stress and radiation [112, 113]. Egr1, as an immediate early-response transcription factor and zinc-finger protein, is expressed in LT-HSCs. $Egr1^{-/-}$ mice show increased proliferation of HSCs, resulting in the mobilization into the blood by mechanisms that are so far unknown. This suggests that Egr1 acts as a retention and quiescence factor in HSCs [114].

In addition to the maintenance of cell cycle quiescence, the regulation of asymmetric cell division represents another critical component for the maintenance of stem cell numbers. During asymmetrical cell division, 1 HSC generates 2 daughter cells; one remains quiescent in the niche, whereas the other differentiates and leaves the niche to produce progenitor cells [115–118]. The regulation of asymmetrical HSC division is highly controversial. One theory is the unequal distribution of cell fate determinants (such as transcription factors, mRNA or microRNAs) into the 2 daughter cells during cell division [119–121]. Another possibility is that the environment influences stem cell fate. Thereby, 1 daughter cell just leaves the niche and differentiates, whereas the residual cell preserves the stem cell phenotype [122, 123]. Although much is known about the signaling pathways and neighboring cells involved in the stem cell self-renewal, the interplay between the different pathway components as well as the spatiotemporal regulation of these mechanisms are still unknown. Dr. Tannishtha Reya developed an approach taking advantage of the Notch reporter mice, which showed high expression of green fluorescent protein in HSCs but diminished expression during differentiation. Dr. Reya used this model to study the developmental signals that regulate stem cell fate and the choice that stem cells make between self-renewal and differentiation. Numb is one of the molecules that were inherited preferentially to the differentiated daughter. Utilizing BCR-ABL or Hoxa9 to generate either chronic- or acute-phase leukemia, Dr. Reya and coworkers found that in chronic phase, cells primarily maintained the balance between asymmetric and symmetric divisions, while in acute phase, the division pattern switched over to being primarily symmetric. Together, these data support the concept that aberrant stem cell fate decisions play an important role in the progression of cancer and could also be a hallmark of aging.

To characterize factors that regulate quiescence of HSCs in response stress, Dr. Margaret A. Goodell and colleagues used knockout mouse models. The mice were treated with 5-FU to induce cycling of LT-HSCs, which was followed by gene expression profiling. Six days after 5-FU treatment, all HSCs were cycling, and gene profiling was conducted to compare gene expression in proliferating and quiescent HSCs. One outstanding molecule was CD48, which is already known as a marker for the differentiation status of HSCs (see above) and as a regulator of T cell development. CD48 is a GPI-linked SLAM family member, which is involved in T cell activation and is expressed in all differentiated blood cells. CD48 is expressed on all transiently activated HSCs [2, 8, 124], which leads to the question whether CD48 has a role in the regulation of HSC quiescence and proliferation.

In transplantation experiments, Dr. Goodell and coworkers showed that HSCs of $CD48^{-/-}$ mice display significant reconstitution defects. But how does a molecule which is not expressed on quiescent HSCs impact the reconstitution capacity? It is noteworthy that the numbers of ST-HSCs are increased in $CD48^{-/-}$ mice, as are the number of MPPs. In contrast, the number of differentiated blood cells is decreased in $CD48^{-/-}$ mice compared to $CD48^{+/+}$ mice. Dr. Goodell presented data to indicate that interactions with other known SLAM molecules, such as CD244, were important in this cell number regulation, and that IFNγ production was perturbed. However, the underlying molecular mechanism remains to be defined. In conclusion, Dr. Goodell suggests that CD48 can act as a sensor of cytokines in the HSC niche and as an indirect regulator of LT-HSCs. According to this model, CD48-mediated induction of IFNγ can suppress the production of cytokines in the HSC niche.

Regulation of Dormant HSCs

In line with a role for IFNs in the control of stem cell quiescence was recent work that Prof. Trumpp presented at the symposium [125]. His data also connected the topics of stem cell quiescence and cancer stem cells. The idea behind this study was the clinical problem of tumor relapse in chronic myeloid leukemia (CML), which occurs years after the remission-inducing treatment with the specific BCR-ABL-kinase-inhibitor 'imatinib' [126].

In over 90% of CML patients, the so-called Philadelphia chromosome can be found. The

Philadelphia chromosome is a shortened version of chromosome 22, generated by a recombination of regions on chromosomes 9 and 22. Via this translocation, the tyrosine kinase ABL on chromosome 9 is fused to a fragment of the BCR gene on chromosome 22. The mutated cells produce a fusion protein, BCR-ABL, that shows enhanced tyrosine kinase activity, which results in the increased proliferation of white blood cells [127, 128]. Imatinib inhibits ABL in the mutated cells by blocking the ATP-binding site, thereby suppressing proliferation of white blood cells. The medicament affects also other kinases, but healthy cells can compensate this blockage by using alternative signaling pathways [129, 130].

Due to the surprising immediate relapse of CML patients present in molecular remission after stopping imatinib treatment, it was postulated that there must be a therapy-resistant CML stem cell. The resistance is likely mediated by the deep quiescence of at least some CML stem cells. These quiescent tumor cells become reactivated causing a relapse of the leukemia. HSCs are partially similar to tumor stem cells in terms of quiescence. Prof. Trumpp and colleagues could show that the LSKCD150^{+}CD34^{low} undifferentiated HSCs are subdivided into 2 populations. There is an activated population, which has entered the cell cycle and is able to reconstitute mice, but not in serial transplantations. In addition, the bone marrow contains a very small, strongly quiescent sub-population of CD34^{low} HSCs, which very rarely enters the cell cycle. These 'dormant' HSCs cycle only 5 times during the lifespan of a mouse and they are mostly responsible for injury repair but not for the daily regeneration of the blood system [131]. Long-term quiescent, dormant HSCs have a reduced metabolism and a silenced replication machinery. Prof. Trumpp and colleagues showed that the dormant HSCs become activated by injury signals, such as 5-FU treatment, and can go back into dormancy after the successful replenishment of the hematopoietic compartment [131].

An interesting task of HSC research is now to define the molecular pathways controlling hibernation of HSCs. There are already some data about the importance of lipid rafts in hibernation. Dr. Nakauchi showed that the PI3K-Akt-FoxO signaling pathway is inactive in freshly isolated HSCs but active in cycling progenitor cells. Thereby, the lipid raft status finely tunes cytokine signal levels and regulates Akt activity. Lipid rafts are cholesterol- and glycosphingolipid-enriched spots in the membrane of the cells containing a variety of functional molecules and acting as platforms for cellular functions. The size of the lipid rafts determines the signal intensity and functional outcome by the formation of clusters. These clusters are absent in quiescent HSCs but can be clearly observed in activated stem cells [132]. The clustering of lipid rafts in turn has been shown to be negatively regulated by TGF-β. TGF-β significantly inhibits cytokine-induced activation of c-Src, one of the lipid raft components, in HSCs. Although the precise molecular mechanism for this inhibition is so far unknown, Src family kinases could be key targets for TGF-β in affecting lipid raft reorganization [133].

Prof. Trumpp reported recently published data showing that treatment with IFNα increases HSC numbers. Interferons, especially IFNα, inhibit virus replication and are strongly produced during viral infection [134–137]. IFNα can directly activate HSCs. This signaling is not mediated by the bone marrow niche but directly activates the dormant HSCs. This was shown by transplantation experiments of wild-type HSCs into IFNRα mice. The transplanted HSCs were fully responsive to IFNα in the recipient mice. Prof. Trumpp and colleagues showed that the stimulatory effects are mediated via the Jak/STAT signaling pathway. In detail, STAT1 is the mediator for the action of IFNα on dormant HSCs (fig. 5). STAT1 activates Sca1 expression, culminating in the proliferation of HSCs. Treatment with IFNα has to be discriminated in acute

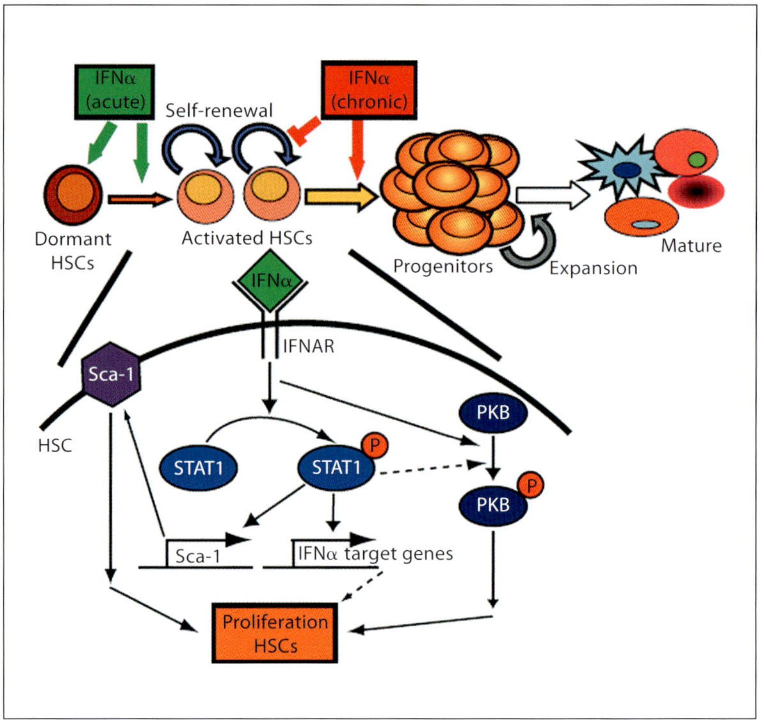

Fig. 5. Overview of the activating effects of IFNα on ST- and LT-HSCs. Acute treatment of mice with IFNα leads to activation of dormant HSCs via activation of STAT1 and secondary up-regulation of Sca1 expression, resulting in proliferation of HSCs. IFNAR = Interferon alpha receptor; PKB = protein kinase B. Reproduced with permission from Essers et al. [125].

and chronic treatment. Acute treatment activates HSCs whereas a chronic treatment leads to the loss of HSCs, at least in competitive transplantation experiments where IFNα-resistant cells overgrow the chronically stimulated cells. To determine whether the dormant HSCs are affected by IFNα, wild-type mice were pretreated with an acute dose of IFNα and later with 5-FU. All mice died due to the complete loss of HSCs. This is not the case when both treatments are given at the same time point.

In summary, IFNα activates dormant HSCs which can be destroyed by a later treatment with 5-FU. A treatment at the same time point would not be effective due to the fact that dormant HSCs have first to be fully activated and will then be sensitive to 5-FU. Prof. Trumpp suggested that IFNα treatment could also activate quiescent CML stem cells that could then be eliminated by imatinib treatment, which may improve treatment results and possibly lead to the elimination of CML stem cells. However, possible side effects may occur. For example, hepatitis patients are treated with interferons against their disease, which could lead in turn to increased sensitivity to cytotoxic agents like 5-FU [125]. It remains to be analyzed whether the here-described role of interferon signaling also influences the aging of HSCs.

Summary

In summary, the above-described studies and meeting presentations suggest that there are age-dependent changes in the molecular pathways of HSCs. Whether these changes are due to intrinsic HSC aging or whether these changes rather reflect the selection for distinct cell populations remains to be investigated. In addition, it will be interesting to study whether these mechanisms inhibit the initiation of cancer or lead to an accelerated selection of malignant HSCs and cancer promotion (see chapter 5). The increasing understanding of molecular changes in aging HSCs will help to develop molecular therapies aiming to improve hematopoiesis in the elderly. In addition, the understanding of HSC aging will contribute to discover prevention strategies and new therapies for the treatment of hemato-lymphopoietic malignancies that show an increasing incidence in older people.

References

1 Ema H, Morita Y, Yamazaki S, Matsubara A, Seita J, Tadokoro Y, Kondo H, Takano H, Nakauchi H: Adult mouse hematopoietic stem cells: purification and single-cell assays. Nat Protoc 2006; 1:2979–2987.

2 Mackey MC: Cell kinetic status of haematopoietic stem cells. Cell Proliferation 2001;34:71–83.

3 Weissman IL, Shizuru JA: The origins of the identification and isolation of hematopoietic stem cells, and their capability to induce donor-specific transplantation tolerance and treat autoimmune diseases. Blood 2008;112: 3543–3553.

4 Bryder D, Rossi DJ, Weissman IL: Hematopoietic stem cells: the paradigmatic tissue-specific stem cell. Am J Pathol 2006;169:338–346.

5 Baum CM, Weissman IL, Tsukamoto AS, Buckle AM, Peault B: Isolation of a candidate human hematopoietic stem-cell population. Proc Natl Acad Sci USA 1992;89:2804–2805.

6 Goodell MA, Brose K, Paradis G, Conner AS, Mulligan RC: Isolation and functional properties of murine hematopoietic stem cells that are replicating in vivo. J Exp Med 1996;183:1797–1806.

7 Goodell MA, Rosenzweig M, Kim H, Marks DF, DeMaria M, Paradis G, Grupp SA, Sieff CA, Mulligan RC, Johnson RP: Dye efflux studies suggest that hematopoietic stem cells expressing low or undetectable levels of CD34 antigen exist in multiple species. Nat Med 1997;3:1337–1345.

8 Kiel MJ, Yilmaz OH, Iwashita T, Terhorst C, Morrison SJ: SLAM family receptors distinguish hematopoietic stem and progenitor cells and reveal endothelial niches for stem cells. Cell 2005; 121:1109–1121.

9 Morrison SJ, Weissman IL: The long-term repopulating subset of hematopoietic stem cells is deterministic and isolatable by phenotype. Immunity 1994;1:661–673.

10 Osawa M, Hanada K, Hamada H, Nakauchi H: Long-term lymphohematopoietic reconstitution by a single CD34-low/negative hematopoietic stem cell. Science 1996;273:242–245.

11 Spangrude GJ, Heimfeld S, Weissman IL: Purification and characterization of mouse hematopoietic stem cells. Science 1988;241:58–62.

12 Akashi K, He X, Chen J, Iwasaki H, Niu C, Steenhard B, Zhang J, Haug J, Li L: Transcriptional accessibility for genes of multiple tissues and hematopoietic lineages is hierarchically controlled during early hematopoiesis. Blood 2003;101:383–389.

13 Lerner C, Harrison DE: 5-Fluorouracil spares hemopoietic stem cells responsible for long-term repopulation. Exp Hematol 1990;18:114–118.

14 Ramalho-Santos M, Yoon S, Matsuzaki Y, Mulligan RC, Melton DA: 'Stemness': transcriptional profiling of embryonic and adult stem cells. Science 2002;298: 597–600.

15 Uchida N, Dykstra B, Lyons KJ, Leung FY, Eaves CJ: Different in vivo repopulating activities of purified hematopoietic stem cells before and after being stimulated to divide in vitro with the same kinetics. Exp Hematol 2003;31: 1338–1347.

16 Venezia TA, Merchant AA, Ramos CA, Whitehouse NL, Young AS, Shaw CA, Goodell MA: Molecular signatures of proliferation and quiescence in hematopoietic stem cells. PLoS Biol 2004; 2:e301:1640–1651.

17 Akashi K, Traver D, Miyamoto T, Weissman IL: A clonogenic common myeloid progenitor that gives rise to all myeloid lineages. Nature 2000;404: 193–197.

18 Kondo M, Weissman IL, Akashi K: Identification of clonogenic common lymphoid progenitors in mouse bone marrow. Cell 1997;91:661–672.

19 Christensen JL, Weissman IL: Flk-2 is a marker in hematopoietic stem cell differentiation: a simple method to isolate long-term stem cells. Proc Natl Acad Sci USA 2001;98:14541–14546.

20 Cho RH, Sieburg HB, Muller-Sieburg CE: A new mechanism for the aging of hematopoietic stem cells: aging changes the clonal composition of the stem cell compartment but not individual stem cells. Blood 2008;111:5553–5561.

21 Muller-Sieburg C, Sieburg HB: Stem cell aging: survival of the laziest? Cell Cycle 2008;7:3798–3804.

22 Czechowicz A, Kraft D, Weissman IL, Bhattacharya D: Efficient transplantation via antibody-based clearance of hematopoietic stem cell niches. Science 2007;318:1296–1299.

23 Morrison SJ, Wandycz AM, Akashi K, Globerson A, Weissman IL: The aging of hematopoietic stem cells. Nat Med 1996;2:1011–1016.

24 Sudo K, Ema H, Morita Y, Nakauchi H: Age-associated characteristics of murine hematopoietic stem cells. J Exp Med 2000;192:1273–1280.

25 Rossi DJ, Jamieson CH, Weissman IL: Stems cells and the pathways to aging and cancer. Cell 2008;132:681–696.

26 Yilmaz OH, Kiel MJ, Morrison SJ: SLAM family markers are conserved among hematopoietic stem cells from old and reconstituted mice and markedly increase their purity. Blood 2006; 107:924–930.

27 Atkinson SP, Keith WN: Epigenetic control of cellular senescence in disease: opportunities for therapeutic intervention. Expert Rev Mol Med 2007;9:1–26.
28 Brena RM, Huang TH, Plass C: Quantitative assessment of DNA methylation: potential applications for disease diagnosis, classification, and prognosis in clinical settings. J Mol Med 2006;84: 365–377.
29 Franke A, DeCamillis M, Zink D, Cheng N, Brock HW, Paro R: Polycomb and polyhomeotic are constituents of a multimeric protein complex in chromatin of *Drosophila melanogaster.* Embo J 1992;11:2941–2950.
30 Paro R: Propagating memory of transcriptional states. Trends Genet 1995; 11:295–297.
31 Pirrotta V: PcG complexes and chromatin silencing. Curr Opin Genet Dev 1997;7:249–258.
32 Shao Z, Raible F, Mollaaghababa R, Guyon JR, Wu CT, Bender W, Kingston RE: Stabilization of chromatin structure by PRC1, a Polycomb complex. Cell 1999;98:37–46.
33 Rada-Iglesias A, Enroth S, Andersson R, Wanders A, Pahlman L, Komorowski J, Wadelius C: Histone H3 lysine 27 trimethylation in adult differentiated colon associated to cancer DNA hypermethylation. Epigenetics 2009;4:107–113.
34 Zhang L, Zhong K, Dai Y, Zhou H: Genome-wide analysis of histone H3 lysine 27 trimethylation by ChIP-chip in gastric cancer patients. J Gastroenterol 2009;44:305–312.
35 Gunster MJ, Satijn DP, Hamer KM, den Blaauwen JL, de Bruijn D, Alkema MJ, van Lohuizen M, van Driel R, Otte AP: Identification and characterization of interactions between the vertebrate polycomb-group protein BMI1 and human homologs of polyhomeotic. Mol Cell Biol 1997;17:2326–2335.
36 Hobert O, Jallal B, Ullrich A: Interaction of Vav with ENX-1, a putative transcriptional regulator of homeobox gene expression. Mol Cell Biol 1996;16: 3066–3073.
37 Laible G, Wolf A, Dorn R, Reuter G, Nislow C, Lebersorger A, Popkin D, Pillus L, Jenuwein T: Mammalian homologues of the Polycomb-group gene Enhancer of zeste mediate gene silencing in *Drosophila* heterochromatin and at *S. cerevisiae* telomeres. Embo J 1997;16:3219–3232.
38 Satijn DP, Gunster MJ, van der Vlag J, Hamer KM, Schul W, Alkema MJ, Saurin AJ, Freemont PS, van Driel R, Otte AP: RING1 is associated with the polycomb group protein complex and acts as a transcriptional repressor. Mol Cell Biol 1997;17:4105–4113.
39 Schumacher A, Faust C, Magnuson T: Positional cloning of a global regulator of anterior-posterior patterning in mice. Nature 1996;383:250–253.
40 Sewalt RG, van der Vlag J, Gunster MJ, Hamer KM, den Blaauwen JL, Satijn DP, Hendrix T, van Driel R, Otte AP: Characterization of interactions between the mammalian polycomb-group proteins Enx1/EZH2 and EED suggests the existence of different mammalian polycomb-group protein complexes. Mol Cell Biol 1998;18:3586–3595.
41 Tagawa M, Sakamoto T, Shigemoto K, Matsubara H, Tamura Y, Ito T, Nakamura I, Okitsu A, Imai K, Taniguchi M: Expression of novel DNA-binding protein with zinc finger structure in various tumor cells. J Biol Chem 1990;265: 20021–20026.
42 van Lohuizen M, Tijms M, Voncken JW, Schumacher A, Magnuson T, Wientjens E: Interaction of mouse polycomb-group (Pc-G) proteins Enx1 and Enx2 with Eed: indication for separate Pc-G complexes. Mol Cell Biol 1998;18: 3572–3579.
43 van Lohuizen M, Verbeek S, Scheijen B, Wientjens E, van der Gulden H, Berns A: Identification of cooperating oncogenes in E mu-myc transgenic mice by provirus tagging. Cell 1991;65:737–752.
44 Bhattacharyya J, Mihara K, Yasunaga S, Tanaka H, Hoshi M, Takihara Y, Kimura A: Bmi-1 expression is enhanced through transcriptional and posttranscriptional regulation during the progression of chronic myeloid leukemia. Ann Hematol 2009;88:333–340.
45 Faubert A, Chagraoui J, Mayotte N, Frechette M, Iscove NN, Humphries RK, Sauvageau G: Complementary and independent function for Hoxb4 and Bmi1 in HSC activity. Cold Spring Harb Symp Quant Biol 2008;73:555–564.
46 Ohtsubo M, Yasunaga S, Ohno Y, Tsumura M, Okada S, Ishikawa N, Shirao K, Kikuchi A, Nishitani H, Kobayashi M, Takihara Y: Polycomb-group complex 1 acts as an E3 ubiquitin ligase for Geminin to sustain hematopoietic stem cell activity. Proc Natl Acad Sci USA 2008;105:10396–10401.
47 Park IK, Qian D, Kiel M, Becker MW, Pihalja M, Weissman IL, Morrison SJ, Clarke MF: Bmi-1 is required for maintenance of adult self-renewing haematopoietic stem cells. Nature 2003;423: 302–305.
48 Kajiume T, Ohno N, Sera Y, Kawahara Y, Yuge L, Kobayashi M: Reciprocal expression of Bmi1 and Mel-18 is associated with functioning of primitive hematopoietic cells. Exp Hematol 2009; 37:857–866.
49 Lessard J, Sauvageau G: Bmi-1 determines the proliferative capacity of normal and leukaemic stem cells. Nature 2003;423:255–260.
50 van der Lugt NM, Domen J, Linders K, van Roon M, Robanus-Maandag E, te Riele H, van der Valk M, Deschamps J, Sofroniew M, van Lohuizen M, Berns A: Posterior transformation, neurological abnormalities, and severe hematopoietic defects in mice with a targeted deletion of the Bmi-1 proto-oncogene. Genes Dev 1994;8:757–769.
51 Molofsky AV, Pardal R, Iwashita T, Park IK, Clarke MF, Morrison SJ: Bmi-1 dependence distinguishes neural stem cell self-renewal from progenitor proliferation. Nature 2003;425:962–967.
52 Pietersen AM, Evers B, Prasad AA, Tanger E, Cornelissen-Steijger P, Jonkers J, van Lohuizen M: Bmi1 regulates stem cells and proliferation and differentiation of committed cells in mammary epithelium. Curr Biol 2008; 18:1094–1099.
53 Bruggeman SW, Valk-Lingbeek ME, van der Stoop PP, Jacobs JJ, Kieboom K, Tanger E, Hulsman D, Leung C, Arsenijevic Y, Marino S, van Lohuizen M: Ink4a and Arf differentially affect cell proliferation and neural stem cell self-renewal in Bmi1-deficient mice. Genes Dev 2005;19:1438–1443.
54 Jacobs JJ, Kieboom K, Marino S, DePinho RA, van Lohuizen M: The oncogene and Polycomb-group gene bmi-1 regulates cell proliferation and senescence through the ink4a locus. Nature 1999; 397:164–168.

55 Molofsky AV, He S, Bydon M, Morrison SJ, Pardal R: Bmi-1 promotes neural stem cell self-renewal and neural development but not mouse growth and survival by repressing the p16Ink4a and p19Arf senescence pathways. Genes Dev 2005;19:1432–1437.
56 Oguro H, Iwama A, Morita Y, Kamijo T, van Lohuizen M, Nakauchi H: Differential impact of Ink4a and Arf on hematopoietic stem cells and their bone marrow microenvironment in Bmi1-deficient mice. J Exp Med 2006;203: 2247–2253.
57 Agherbi H, Gaussmann-Wenger A, Verthuy C, Chasson L, Serrano M, Djabali M: Polycomb mediated epigenetic silencing and replication: timing at the Ink4a/Arf locus during senescence. PLoS One 2009;e5622:1–10.
58 Ringrose L, Paro R: Epigenetic regulation of cellular memory by the Polycomb and Trithorax group proteins. Annu Rev Genet 2004;38:413–443.
59 Hanson RD, Hess JL, Yu BD, Ernst P, van Lohuizen M, Berns A, van der Lugt NM, Shashikant CS, Ruddle FH, Seto M, Korsmeyer SJ: Mammalian Trithorax and polycomb-group homologues are antagonistic regulators of homeotic development. Proc Natl Acad Sci USA 1999;96:14372–14377.
60 Jacobs JJ, van Lohuizen M: Polycomb repression: from cellular memory to cellular proliferation and cancer. Biochim Biophys Acta 2002;1602:151–161.
61 Lund AH, van Lohuizen M: Epigenetics and cancer. Genes Dev 2004;18:2315–2335.
62 Madan V, Madan B, Brykczynska U, Zilbermann F, Hogeveen K, Dohner K, Dohner H, Weber O, Blum C, Rodewald HR, Sassone-Corsi P, Peters AH, Fehling HJ: Impaired function of primitive hematopoietic cells in mice lacking the Mixed-Lineage-Leukemia homolog MLL5. Blood 2009;113:1444–1454.
63 Lombard DB, Chua KF, Mostoslavsky R, Franco S, Gostissa M, Alt FW: DNA repair, genome stability, and aging. Cell 2005;120:497–512.
64 Navarro S, Meza NW, Quintana-Bustamante O, Casado JA, Jacome A, McAllister K, Puerto S, Surralles J, Segovia JC, Bueren JA: Hematopoietic dysfunction in a mouse model for Fanconi anemia group D1. Mol Ther 2006;14:525–535.
65 Nijnik A, Woodbine L, Marchetti C, Dawson S, Lambe T, Liu C, Rodrigues NP, Crockford TL, Cabuy E, Vindigni A, Enver T, Bell JI, Slijepcevic P, Goodnow CC, Jeggo PA, Cornall RJ: DNA repair is limiting for haematopoietic stem cells during ageing. Nature 2007; 447:686–690.
66 Reese JS, Liu L, Gerson SL: Repopulating defect of mismatch repair-deficient hematopoietic stem cells. Blood 2003; 102:1626–1633.
67 Rossi DJ, Bryder D, Seita J, Nussenzweig A, Hoeijmakers J, Weissman IL: Deficiencies in DNA damage repair limit the function of haematopoietic stem cells with age. Nature 2007;447: 725–729.
68 Janzen V, Forkert R, Fleming HE, Saito Y, Waring MT, Dombkowski DM, Cheng T, DePinho RA, Sharpless NE, Scadden DT: Stem-cell ageing modified by the cyclin-dependent kinase inhibitor p16INK4a. Nature 2006;443:421–426.
69 Krishnamurthy J, Ramsey MR, Ligon KL, Torrice C, Koh A, Bonner-Weir S, Sharpless NE: p16INK4a induces an age-dependent decline in islet regenerative potential. Nature 2006;443: 453–457.
70 Molofsky AV, Slutsky SG, Joseph NM, He S, Pardal R, Krishnamurthy J, Sharpless NE, Morrison SJ: Increasing p16INK4a expression decreases forebrain progenitors and neurogenesis during ageing. Nature 2006;443:448–452.
71 Collado M, Blasco MA, Serrano M: Cellular senescence in cancer and aging. Cell 2007;130:223–233.
72 Finkel T, Serrano M, Blasco MA: The common biology of cancer and ageing. Nature 2007;448:767–774.
73 Blasco MA, Lee HW, Hande MP, Samper E, Lansdorp PM, DePinho RA, Greider CW: Telomere shortening and tumor formation by mouse cells lacking telomerase RNA. Cell 1997;91:25–34.
74 Lee HW, Blasco MA, Gottlieb GJ, Horner JW 2nd, Greider CW, DePinho RA: Essential role of mouse telomerase in highly proliferative organs. Nature 1998;392:569–574.
75 Rudolph KL, Chang S, Lee HW, Blasco M, Gottlieb GJ, Greider C, DePinho RA: Longevity, stress response, and cancer in aging telomerase-deficient mice. Cell 1999;96:701–712.
76 Choudhury AR, Ju Z, Djojosubroto MW, Schienke A, Lechel A, Schaetzlein S, Jiang H, Stepczynska A, Wang C, Buer J, Lee HW, von Zglinicki T, Ganser A, Schirmacher P, Nakauchi H, Rudolph KL: Cdkn1a deletion improves stem cell function and lifespan of mice with dysfunctional telomeres without accelerating cancer formation. Nat Genet 2007;39:99–105.
77 Ju Z, Jiang H, Jaworski M, Rathinam C, Gompf A, Klein C, Trumpp A, Rudolph KL: Telomere dysfunction induces environmental alterations limiting hematopoietic stem cell function and engraftment. Nat Med 2007;13:742–747.
78 Samper E, Fernandez P, Eguia R, Martin-Rivera L, Bernad A, Blasco MA, Aracil M: Long-term repopulating ability of telomerase-deficient murine hematopoietic stem cells. Blood 2002;99: 2767–2775.
79 Allsopp RC, Cheshier S, Weissman IL: Telomere shortening accompanies increased cell cycle activity during serial transplantation of hematopoietic stem cells. J Exp Med 2001;193:917–924.
80 Siegl-Cachedenier I, Munoz P, Flores JM, Klatt P, Blasco MA: Deficient mismatch repair improves organismal fitness and survival of mice with dysfunctional telomeres. Genes Dev 2007; 21:2234–2247.
81 Begus-Nahrmann Y, Lechel A, Obenauf AC, Nalapareddy K, Peit E, Hoffmann E, Schlaudraff F, Liss B, Schirmacher P, Kestler H, Danenberg E, Barker N, Clevers H, Speicher MR, Rudolph KL: A genetic and genomic analysis identifies a cluster of genes associated with hematopoietic cell turnover. Nat Genet 2009;41:1138–1144.
82 Wu WS, Heinrichs S, Xu D, Garrison SP, Zambetti GP, Adams JM, Look AT: Slug antagonizes p53-mediated apoptosis of hematopoietic progenitors by repressing puma. Cell 2005;123:641–653.
83 Qiu W, Carson-Walter EB, Liu H, Epperly M, Greenberger JS, Zambetti GP, Zhang L, Yu J: PUMA regulates intestinal progenitor cell radiosensitivity and gastrointestinal syndrome. Cell Stem Cell 2008;2:576–583.
84 Bailey DW: Recombinant-inbred strains: an aid to finding identity, linkage, and function of histocompatibility and other genes. Transplantation 1971; 11:325–327.

85 De Haan G, Van Zant G: Genetic analysis of hemopoietic cell cycling in mice suggests its involvement in organismal life span. Faseb J 1999;13:707–713.
86 Geiger H, True JM, de Haan G, Van Zant G: Age- and stage-specific regulation patterns in the hematopoietic stem cell hierarchy. Blood 2001;98:2966–2972.
87 Henckaerts E, Langer JC, Orenstein J, Snoeck HW: The positive regulatory effect of TGF-beta2 on primitive murine hemopoietic stem and progenitor cells is dependent on age, genetic background, and serum factors. J Immunol 2004;173:2486–2493.
88 Broman KW: Mapping expression in randomized rodent genomes. Nat Genet 2005;37:209–210.
89 de Haan G, Van Zant G: Intrinsic and extrinsic control of hematopoietic stem cell numbers: mapping of a stem cell gene. J Exp Med 1997;186:529.
90 Muller-Sieburg C, Riblet R: Genetic control of the frequency of hematopoietic stem cells in mice: mapping of a candidate locus to chromosome. J Exp Med 1996;183:1141–1150.
91 Kamminga LM, Van Os R, Ausema A, Noach EJK, Weersing E, Dontje B, Vellenga E, De Haan G: Impaired hematopoietic stem cell functioning after serial transplantation and during normal aging. Stem Cells 2005;23:82–92.
92 Vanderluit JL, Wylie CA, McClellan KA, Ghanem N, Fortin A, Callaghan S, MacLaurin JG, Park DS, Slack RS: The retinoblastoma family member p107 regulates the rate of progenitor commitment to a neuronal fate. J Cell Biol 2007;178:129–139.
93 Stevaux O, Dimova D, Frolov MV, Taylor-Harding B, Morris E, Dyson N: Distinct mechanisms of E2F regulation by *Drosophila* RBF1 and RBF2. Embo J 2002;21:4927–4937.
94 Stevaux O, Dyson NJ: A revised picture of the E2F transcriptional network and RB function. Curr Opin Cell Biol 2002;14:684–691.
95 Vanderluit JL, Ferguson KL, Nikoletopoulou V, Parker M, Ruzhynsky V, Alexson T, McNamara SM, Park DS, Rudnicki M, Slack RS: p107 regulates neural precursor cells in the mammalian brain. J Cell Biol 2004;166:853–863.
96 Ho KK, Myatt SS, Lam EW: Many forks in the path: cycling with FoxO. Oncogene 2008;27:2300–2311.
97 Markel P, Shu P, Ebeling C, Carlson GA, Nagle DL, Smutko JS, Moore KJ: Theoretical and empirical issues for marker-assisted breeding of congenic mouse strains. Nat Genet 1997;17:280–284.
98 Wakeland E, Morel L, Achey K, Yui M, Longmate J: Speed congenics: a classic technique in the fast lane (relatively speaking). Immunol Today 1997;18:472–477.
99 Liang Y, Jansen M, Aronow B, Geiger H, Van Zant G: The quantitative trait gene latexin influences the size of the hematopoietic stem cell population in mice. Nat Genet 2007;39:178–188.
100 Cheng T, Rodrigues N, Shen H, Yang Y, Dombkowski D, Sykes M, Scadden DT: Hematopoietic stem cell quiescence maintained by p21cip1/waf1. Science 2000;287:1804–1808.
101 van Os R, Kamminga LM, Ausema A, Bystrykh LV, Draijer DP, van Pelt K, Dontje B, de Haan G: A limited role for p21Cip1/Waf1 in maintaining normal hematopoietic stem cell functioning. Stem Cells 2007;25:836–843.
102 Dunwell TL, Dickinson RE, Stankovic T, Dallol A, Weston V, Austen B, Catchpoole D, Maher ER, Latif F: Frequent epigenetic inactivation of the SLIT2 gene in chronic and acute lymphocytic leukemia. Epigenetics 2009;4:265–269.
103 Shibata F, Goto-Koshino Y, Morikawa Y, Komori T, Ito M, Fukuchi Y, Houchins JP, Tsang M, Li DY, Kitamura T, Nakajima H: Roundabout 4 is expressed on hematopoietic stem cells and potentially involved in the niche-mediated regulation of the side population phenotype. Stem Cells 2009;27:183–190.
104 Gerrits A, Li Y, Tesson BM, Bystrykh LV, Weersing E, Ausema A, Dontje B, Wang X, Breitling R, Jansen RC, De Haan G: Expression quantitative trait loci are highly sensitive to cellular differentiation state. PLoS Genetics 2009;e1000692:1–8.
105 Bee T, Ashley E, Bickley S, Jarrat A, Li P-S, Sloane-Stanley J, Göttgens B, De Bruijn M: The mouse Runx1 23 hematopoietic stem cell enhancer confers hematopoietic specificity to both Runx1 promoters. Blood 2009;113:5121–5124.
106 Mohn F, Schübeler D: Genetics and epigenetics: stability and plasticity during cellular differentiation. Trends Genet 2009;25:129–136.
107 Yilmaz OH, Valdez R, Theisen BK, Guo W, Ferguson DO, Wu H, Morrison SJ: Pten dependence distinguishes haematopoietic stem cells from leukaemia-initiating cells. Nature 2006;441:475–482.
108 Zhang J, Grindley JC, Yin T, Jayasinghe S, He XC, Ross JT, Haug JS, Rupp D, Porter-Westpfahl KS, Wiedemann LM, Wu H, Li L: PTEN maintains haematopoietic stem cells and acts in lineage choice and leukaemia prevention. Nature 2006;441:518–522.
109 Wilson A, Murphy MJ, Oskarsson T, Kaloulis K, Bettess MD, Oser GM, Pasche AC, Knabenhans C, Macdonald HR, Trumpp A: c-Myc controls the balance between hematopoietic stem cell self-renewal and differentiation. Genes Dev 2004;18:2747–2763.
110 Yuan Y, Shen H, Franklin DS, Scadden DT, Cheng T: In vivo self-renewing divisions of haematopoietic stem cells are increased in the absence of the early G1-phase inhibitor, p18INK4C. Nat Cell Biol 2004;6:436–442.
111 Yu H, Yuan Y, Shen H, Cheng T: Hematopoietic stem cell exhaustion impacted by p18 INK4C and p21 Cip1/Waf1 in opposite manners. Blood 2006;107:1200–1206.
112 Lacorazza HD, Nimer SD: The emerging role of the myeloid Elf-1 like transcription factor in hematopoiesis. Blood Cells Mol Dis 2003;31:342–350.
113 Lacorazza HD, Yamada T, Liu Y, Miyata Y, Sivina M, Nunes J, Nimer SD: The transcription factor MEF/ELF4 regulates the quiescence of primitive hematopoietic cells. Cancer Cell 2006;9:175–187.
114 Min IM, Pietramaggiori G, Kim FS, Passegue E, Stevenson KE, Wagers AJ: The transcription factor EGR1 controls both the proliferation and localization of hematopoietic stem cells. Cell Stem Cell 2008;2:380–391.
115 Brummendorf TH, Dragowska W, Zijlmans J, Thornbury G, Lansdorp PM: Asymmetric cell divisions sustain long-term hematopoiesis from single-sorted human fetal liver cells. J Exp Med 1998;188:1117–1224.

116 Giebel B, Zhang T, Beckmann J, Spanholtz J, Wernet P, Ho AD, Punzel M: Primitive human hematopoietic cells give rise to differentially specified daughter cells upon their initial cell division. Blood 2006;107:2146–2152.
117 Huang S, Law P, Francis K, Palsson BO, Ho AD: Symmetry of initial cell divisions among primitive hematopoietic progenitors is independent of ontogenic age and regulatory molecules. Blood 1999;94:2595–2604.
118 Mayani H, Dragowska W, Lansdorp PM: Lineage commitment in human hemopoiesis involves asymmetric cell division of multipotent progenitors and does not appear to be influenced by cytokines. J Cell Physiol 1993;157: 579–586.
119 Ho AD: Kinetics and symmetry of divisions of hematopoietic stem cells. Exp Hematol 2005;33:1–8.
120 Suda J, Suda T, Ogawa M: Analysis of differentiation of mouse hemopoietic stem cells in culture by sequential replating of paired progenitors. Blood 1984;64:393–399.
121 Takano H, Ema H, Sudo K, Nakauchi H: Asymmetric division and lineage commitment at the level of hematopoietic stem cells: inference from differentiation in daughter cell and granddaughter cell pairs. J Exp Med 2004;199:295–302.
122 Ohlstein B, Kai T, Decotto E, Spradling A: The stem cell niche: theme and variations. Curr Opin Cell Biol 2004;16:693–699.
123 Spradling A, Drummond-Barbosa D, Kai T: Stem cells find their niche. Nature 2001;414:98–104.
124 Gonzalez-Cabrero J, Wise CJ, Latchman Y, Freeman GJ, Sharpe AH, Reiser H: CD48-deficient mice have a pronounced defect in CD4(+) T cell activation. Proc Natl Acad Sci USA 1999;96:1019–1023.
125 Essers MA, Offner S, Blanco-Bose WE, Waibler Z, Kalinke U, Duchosal MA, Trumpp A: IFNalpha activates dormant haematopoietic stem cells in vivo. Nature 2009;458:904–908.
126 Gambacorti-Passerini C, le Coutre P, Mologni L, Fanelli M, Bertazzoli C, Marchesi E, Di Nicola M, Biondi A, Corneo GM, Belotti D, Pogliani E, Lydon NB: Inhibition of the ABL kinase activity blocks the proliferation of BCR/ABL+ leukemic cells and induces apoptosis. Blood Cells Mol Dis 1997;23:380–394.
127 Groffen J, Heisterkamp N: The BCR/ABL hybrid gene. Baillieres Clin Haematol 1987;1:983–999.
128 Wahrman J, Voss R, Shapiro T, Ashkenazi A: The Philadelphia chromosome in two children with chronic myeloid leukemia. Isr J Med Sci 1967; 3:380–391.
129 Deininger M, Buchdunger E, Druker BJ: The development of imatinib as a therapeutic agent for chronic myeloid leukemia. Blood 2005;105:2640–2653.
130 O'Brien SG, Guilhot F, Larson RA, Gathmann I, Baccarani M, Cervantes F, Cornelissen JJ, Fischer T, Hochhaus A, Hughes T, Lechner K, Nielsen JL, Rousselot P, Reiffers J, Saglio G, Shepherd J, Simonsson B, Gratwohl A, Goldman JM, Kantarjian H, Taylor K, Verhoef G, Bolton AE, Capdeville R, Druker BJ: Imatinib compared with interferon and low-dose cytarabine for newly diagnosed chronic-phase chronic myeloid leukemia. N Engl J Med 2003;348:994–1004.
131 Wilson A, Laurenti E, Oser G, van der Wath RC, Blanco-Bose W, Jaworski M, Offner S, Dunant CF, Eshkind L, Bockamp E, Lio P, Macdonald HR, Trumpp A: Hematopoietic stem cells reversibly switch from dormancy to self-renewal during homeostasis and repair. Cell 2008;135:1118–1129.
132 Yamazaki S, Iwama A, Takayanagi S, Morita Y, Eto K, Ema H, Nakauchi H: Cytokine signals modulated via lipid rafts mimic niche signals and induce hibernation in hematopoietic stem cells. Embo J 2006;25:3515–3525.
133 Yamazaki S, Iwama A, Takayanagi S, Eto K, Ema H, Nakauchi H: TGF-beta as a candidate bone marrow niche signal to induce hematopoietic stem cell hibernation. Blood 2009;113: 1250–1256.
134 Borden EC, Sen GC, Uze G, Silverman RH, Ransohoff RM, Foster GR, Stark GR: Interferons at age 50: past, current and future impact on biomedicine. Nat Rev Drug Discov 2007;6: 975–990.
135 Darnell JE Jr, Kerr IM, Stark GR: Jak-STAT pathways and transcriptional activation in response to IFNs and other extracellular signaling proteins. Science 1994;264:1415–1421.
136 Kujawski LA, Talpaz M: The role of interferon-alpha in the treatment of chronic myeloid leukemia. Cytokine Growth Factor Rev 2007;18:459–471.
137 Stark GR, Kerr IM, Williams BR, Silverman RH, Schreiber RD: How cells respond to interferons. Annu Rev Biochem 1998;67:227–264.

Dr. Anett Illing
Institute of Molecular Medicine and
Max Planck Research Department on Stem Cell Aging, University of Ulm
Albert-Einstein-Allee 11
DE–89081 Ulm (Germany)
Tel. +49 731 503 6139, Fax +49 731 503 6102
E-Mail anett.illing@uni-ulm.de

Chapter 3

Rudolph KL (ed): Molecular Mechanisms of Adult Stem Cell Aging.
Else Kröner-Fresenius Symp. Basel, Karger, 2010, vol 1, pp 38–45

Alterations of Systemic Environment Impair Stem Cell Function during Aging

Zhangfa Song[a] · Zhenyu Ju[b]

[a]Institute of Molecular Medicine and Max Planck Research Group on Stem Cell Aging, University of Ulm, Ulm, Germany;
[b]Max-Planck-Partner Group on Stem Cell Aging, Key Laboratory of Human Diseases Comparative Medicine, Ministry of Health, Institute of Laboratory Animal Sciences, Chinese Academy of Medical Sciences, Beijing, PR China

Abstract

Aging is associated with a decline in tissue repair and organ regeneration. There is growing evidence that age-related decline of stem cell function contributes to the failure of organ maintenance and homeostasis, yet it remains to be investigated to what extent this is responsible for the development of age-related pathologies across different organs. In humans, it has been shown that both donor age and recipient age influence patient outcomes in response to hematopoietic stem cell transplantation. In transplantation experiments in mice, both stem cell-intrinsic and -extrinsic defects have been implicated in the age-dependent decline of functional stem cell reserves. During the Else Kröner-Fresenius Symposium on the Molecular Mechanisms of Adult Stem Cell Aging, several speakers indicated that cell-extrinsic cues play an important role in the function and fate decision of stem cells during aging. Extrinsic factors that can influence stem cell function can be divided into 2 compartments: the cellular niche (microenvironment) and the systemic environment (macroenvironment). Evidence from various model systems demonstrates that both stem cell niche and systemic environment are crucial for the maintenance of functional stem cells during aging. Age-associated alterations in the microenvironment limit stem cell function in various stem cell compartments, such as melanocyte stem cells, intestinal stem cells, spermatogonia, muscle stem cells, and bone marrow stem cells. The impact of the stem microenvironment on stem cell function and maintenance during aging will be discussed by Dr. Geiger in chapter 4. Here, we focus on new concepts presented during the meeting regarding the impact of age-dependent, systemic environmental alterations on stem cell aging.

Introduction

Aging is typically associated with a functional decline in many tissues and organs, including impairment of immune function, muscle atrophy, skin wrinkling, bone fragility, decreased cognitive function, cardiac dysfunction, and increased cancer incidence. One common property shared by many of these age-related conditions is a deregulation of tissue homeostasis and a deficit of tissue repair, suggesting that age-related decline in the function of tissue-specific stem cells may contribute – at least in part – to some of these phenotypes of aging. Different studies have identified age-related declines in various stem cell compartments including hematopoietic stem cells (HSCs), skeletal muscle progenitor/stem cells (SMPs), neuronal stem cells and oligodendrocyte precursors, germline stem cells and melanocyte stem cells. Due to the fact that HSCs and SMPs are 2 well-defined tissue-specific stem cells, we will focus on these 2 tissue-specific stem cell compartments in the following sections.

Aging causes changes in the frequency and function of tissue-specific stem cells in both the hematopoietic system and skeletal muscle, which could contribute to the evolution of age-related diseases and disabilities, such as immune deficiency, myeloid cancer and muscle atrophy. In both systems, an impaired regenerative capacity and a growing imbalance in lineage differentiation have been associated with stem cell dysfunction during aging. It has been suggested that some of the stem cell defects are transplantable, indicating that these alterations are in part due to cell-autonomous genetic or epigenetic modifications. However, it is still not clear to what extent stem cell-intrinsic changes contribute to the age-dependent deregulation of organ homeostasis and tissue repair. In human hematopoietic stem cell transplantations, both donor age and recipient age influence patient outcomes [1–4]. In mouse experiments, both stem cell intrinsic and extrinsic defects have been implicated in the age-dependent decline of functional stem cell reserves [5]. In addition to hematopoietic stem cells, other stem cell compartments also showed age-associated alternations in microenvironment limiting stem cell function [6–11]. In line with these observations, additional experimental evidence reported during the Else Kröner-Fresenius Symposium on the Molecular Mechanisms of Adult Stem Cell Aging suggested that some of the stem cell-autonomous alterations could be induced by stem cell-extrinsic cues that change with age. In addition, the coordinated decline in stem cell function in multiple tissues in aged mice argues that age-related stem cell dysfunction in different tissues might be induced by the same alterations in systemic regulators influencing stem cell function. The original study from Dr. Thomas A. Rando and colleagues suggested that aged stem cells can be rejuvenated by circulating factors that are present in the serum of young mice [12] (see also below). In addition, the negative factors accumulate in the serum of aged mice since young stem cells showed a decline in function when exposed to the environment from aged parabiotic connected mice. During the symposium, Dr. Rando reported results from follow-up studies showing that some of the molecular and cellular profiles of aged SMPs are induced by changes in Wnt-dependent transcription (see chapters 4 and 7).

Macroenvironmental Alterations Regulate HSC Function

Impairment in lymphopoiesis is a major phenotype of human aging, which contributes to the reduction in immune function in the elderly [13, 14]. In wild-type mice, a similar skewing of hematopoiesis (decreased lymphopoiesis and increased myelopoiesis) occurs with aging [15, 16]. It has been hypothesized that age-related hematological defects have their origin in cell-intrinsic changes in the function of hematopoietic stem and progenitor cells [16, 17]. Recent data suggest constant communication between old HSCs and local young environment that could change the HSCs gradually over time. Transplantation of young HSCs into irradiated aged wild-type mice resulted in reduced function of HSCs [18]. These data suggested that the myeloid-biased features of aged HSCs were determined by both cell-intrinsic and cell-extrinsic cues.

During the symposium, Dr. Christa Muller-Sieburg showed data indicating that myeloid-biased HSCs cannot become lymphoid-biased HSCs during serial transplantation. In contrast, Dr. Hiromitsu Nakauchi showed some data suggesting that myeloid-biased HSCs can generate lymphoid production after 2 rounds of transplantation. The difference between these studies lies in their experimental design. Dr. Muller-Sieburg used total bone marrow as donor cells for limiting dilution assay, while Dr. Nakauchi used prospectively isolated single HSCs as donors. Dr. Irving L. Weissman commented on these experiments that it is possible that irradiation itself produces an abnormal environment that might promote selection of myeloid-biased HSCs (see chapter 2).

To avoid an irradiation effect on the environment, some researchers have used parabiosis (connecting the blood flow of 2 mice), a unique and powerful model system that can assay for systemic regulators of stem cell function in an in vivo system. The surgical joining of 2 animals generates a cross-circulation in parabiotic mice, which is established within 2–3 days after operation. The chimerism of blood cells in the parabiotic connected mice reaches 50% within 7–10 days. Heterochronic parabiosis (connecting an aged mouse to a young mouse) results in rapid exchange of both soluble factors and cells across the vascular junction, allowing each young and old animal to be exposed to the systemic environment of its partner. The origin of hematopoietic cells can be tracked by using congenic markers CD45.1 and CD45.2.

During the symposium, Dr. Amy J. Wagers presented new data showing that in wild-type mice, the systemic environment has an important role in impairing HSC function during aging. The bias of aged HSCs towards myeloid differentiation was rescued to a balanced production of both myeloid cells and lymphoid cells when they were exposed to a young circulatory system. Moreover, the frequency, number, and repopulating capacity of hematopoietic stem cells in aged mice can be rejuvenated by exposure to a young environment. The most intriguing part of these experiments is that a young circulatory system also recovered the osteoblastic niche cell frequency, number and function. Together, these data place the alteration of systemic environment at the top of a series of events that contribute to the decline in function of HSCs during aging.

However, there are still some arguments about the parabiosis model. One of the biggest problems is the cross-engraftment of hematopoietic cells, which could contribute to the local environment in a paracrine manner. In addition, this model could induce stress responses that may trigger secretion of cytokines that influence the HSC function. The level of stress of heterochronic connected mice (young and old) could be different from that of isochronic pairs (young-young or old-old).

Recently, Dr. Tao Cheng's lab showed the leukemia environment can influence the function of HSCs and progenitor cells. Using a leukemic mouse model induced by over-expression of *Notch1*, they studied the kinetics and functions of hematopoietic stem and progenitor cells. In general, hematopoiesis was progressively suppressed during leukemia development. However, a leukemic environment impacted hematopoietic stem cells differently from progenitor cells. The non-transformed HSCs in leukemic mice were kept in a more quiescent state compared to HSCs from non-leukemic mice but remained highly functional upon transplantation to non-leukemic recipients. In contrast, non-transformed hematopoietic progenitor cells in leukemic mice demonstrated transient accelerated proliferation and exhaustion. This phenotype correlated with changes of cell cycle-regulating genes (such as p21, p27 and p18) that are known to control proliferation and quiescence of HSCs [19]. Together, these data support the concept that extrinsic factors play an important role in regulating the fate and function of HSCs.

This concept was further strengthened by observations in cross-transplantation experiments using telomere-dysfunctional mice. Telomeres are composed of tandem double-stranded DNA repeats and associated telomeric binding proteins. The main function of telomeres is to cap the ends of chromosomes and protect them from end-to-end fusion and to avoid the evolution of chromosomal instability. Telomeres shorten with each round of cell division due to the end replication problem of DNA polymerase and due to processing of telomeres during the cell cycle. The enzyme telomerase is required to compensate telomere shorting. Mice lacking telomerase showed progressive telomere shortening and a premature aging phenotype, especially in highly proliferative organs, such as the skin, gut and hematopoietic systems [20–22]. One of the strong aging phenotypes of late-generation telomerase-deficient mice is the skewing in lympho-myelopoiesis [23]. This phenotype was mainly attributed to cell-extrinsic alterations induced by telomere dysfunction. Skewing of hemato-lymphopoiesis was rescued when telomere-dysfunctional HSCs were transplanted into a young wild-type environment. Young wild-type HSCs showed skewing when transplanted into aged telomere-dysfunctional recipients [23]. The relative contribution of telomere dysfunction-induced alterations in the macroenvironment, compared to alterations in the stem cell niche, to the development of skewing in hemato-lymphopoiesis of telomere-dysfunctional mice remains to be determined.

During the meeting, Dr. Zhenyu Ju presented new data on bone transplantation from Dr. Karl Lenhard Rudolph's group addressing this question. Interestingly, a strong reduction in B lymphopoiesis and an increase in myelopoiesis were observed in the grafted bones in the kidney capsule of telomere-dysfunctional recipients compared to those in wild-type recipients, independent of the genotype of the grafted bones (derived from wild-type or telomere-dysfunctional mice). These findings indicate that the systemic environment rather than the niche has a dominant role in impairing lymphopoiesis in response to telomere dysfunction [24].

Macroenvironmental Alterations Regulate SMP Function

SMPs represent another well-studied stem cell compartment, which can be prospectively isolated by FACS using surface markers ($CD45^-$, $Mac\text{-}1^-$, $Sca\text{-}1^-$, $b1\text{-}intergrin^+$, $CXCR4^+$) [25–27] (see chapters 4 and 7). These cells express satellite cell markers ($Pax7^+$), but no differentiation markers (MyoD, Desmin, MyHC). SMPs are committed myogenic progenitor cells, and can reconstitute dystrophic muscle in mdx mice after transplantation [28] (see chapter 4). Age-related skeletal muscle dystrophy is associated with a decline in stem cell number and function showing a decreased myogenic potential but an increase in fibrogenic or adipogenic differentiation (see chapters 4 and 7). A number of studies have shown that cell-extrinsic mechanisms are involved in the functional decline of aging SMPs. Heterochronic parabiosis (see above) significantly restored the function of SMPs and muscle regenerative activity [12]. Old isochronic parabionts are unable to efficiently repair muscle damage, whereas old animals joined to young partners show improved repair capacity. This phenomenon relates not to cross-engraftment of young muscle stem cells from the young partner but to specific changes in the SMPs from the old mice. Dr. Rando reported that this rejuvenation effect in the muscle is stable, and can be maintained for several weeks after separation of a previously parabiosed old mouse. In vitro experiments have shown that young cells exposed to old serum have decreased potential to develop a myogenic phenotype and lose myogenic lineage. In contrast, old cells exposed to young serum showed an improvement in regenerative myogenesis [12]. Together, these data support the concept that systemic environmental alterations play an important role in suppressing muscle stem cell activity in the aged mice by acting perhaps directly on the stem cell or stem cell niche. Recent work of the Rando laboratory indicated that increased expression of ligands that activate the Wnt pathway contributes to the systemic alterations limiting SMPs function during aging (see chapter 4).

Macroenvironmental Aging Characterized by Alterations of Circulatory Factors

In humans, circulating factors including cytokines, chemokines, growth factors and angiogenic factors show differences in expression with aging. For instance, serum IFNγ-inducible chemokines (MIG and IP-10), eotaxin, chemoatractant for eosinophils, and soluble TNF receptor II increase with advancing age, while serum levels of epithelial growth factor receptor and epithelial growth factor decreased with age [29]. These secreted factors may involve regulating the stem cell function during aging. In the hematopoietic system, an age-dependent increase of various cytokines and growth factors in blood serum has been implicated in the stem cell dysfunction in telomere-dysfunctional mice [23]. One of the upregulated factors is granulocyte colony-stimulating factor (G-CSF) which showed an age-dependent increase in telomere-dysfunctional mice but not in wild-type mice. Treatment with human recombinant G-CSF in young wild-type mice mimicked some of the phenotypes of aged telomere-dysfunctional mice, including impairment of B cell development in bone marrow as well as reduced engraftment of transplanted HSCs. In contrast, application of neutralizing antibodies lowered the endogenous G-CSF level in the plasma of aged mice and improved HSC function and engraftment [23]. A recent study demonstrated that extrinsic physiological cytokines (macrophage colony-stimulating factor as well as G-CSF) can instruct hematopoietic lineage choice [30]. Another study showed that thrombopoietin (THPO) modulates HSC cell cycle progression at the osteoblast surface by the THPO/MPL pathway [31]. Treatment with THPO increased expression of cyclin-dependent kinase inhibitors in HSCs and this was associated with maintaining quiescent HSCs. In contrast, treatment with neutralizing anti-Mpl antibody releases native HSCs from their microniches, allowing their displacement by transplanted HSCs [31, 32]. This pathway can also be triggered by other regulators such as LNK, which interacts with JAK2, implicating the signaling axis of THPO/MPL/JAK2/LNK as a regulatory pathway in the control of HSC self-renewal and quiescence [33].

During the symposium, Dr. Andreas Trumpp and colleagues showed that IFNα can stimulate turnover and proliferation, thus leading to exhaustion of HSCs in vivo [34, 35]. In response to treatment with IFNα, HSCs show enhanced exit from G_0 and enter the cell cycle. These data indicate that alterations of cytokine profiles have an impact on the maintenance and function of HSCs (see chapter 2). It remains to be determined which cytokines contribute to the impairment of stem cell function during aging.

DNA Damage and Telomere Dysfunction Induce Age-Dependent Secretion of Circulating Factors

The above experimental data indicate that age-dependent systemic environmental alterations limit stem cell function through abnormal secretion of positive or negative factors that alter signaling pathways influencing the self-renewal and lineage potential of stem cells. However, the underlying mechanisms of this age-dependent secretory phenotype remain unclear. The cellular sources (cell types and tissues) of these circulating factors also remain elusive. One possible mechanism could be the induction of a secretory phenotype in aged (senescent) cells by accumulation of DNA damage. Experimental data have shown that senescent cells develop a complex senescence-associated secretory phenotype in cell culture experiments [36–39]. Among these factors, IL-6 and IL-8, 2 classical NF-κB-responsive cytokines, are associated with normal healing in damaged tissues as well as age-related phenotypic changes, particularly those that resemble chronic inflammatory disease, such as decreased lean body mass, low-grade anemia, multiple myeloma, osteoporosis and Alzheimer's disease [40–42]. More recent data showed persistent DNA damage signaling triggers senescence-associated inflammatory cytokine secretion, including IL-6. These data demonstrated a novel role of the DNA damage responses in aged tissues influencing the cellular environment [43]. Although the secretory phenotype seems to be originated from fully differentiated senescent cells, the possibility cannot be excluded that damaged stem cells might also create a diseased microenvironment in a paracrine manner. Indeed, mesenchymal stem cells can secret cytokines that promote proliferation of hematopoietic progenitors [44]. Increased DNA damage signaling has been documented in aged wild-type mice and is more pronounced in telomere-dysfunctional mice [16, 45, 46]. Jiang et al. [46] have identified a set of 4 bio-markers in aged telomere-dysfunctional mice. These markers were exclusively up-regulated in aged telomere-dysfunctional mice but not in aged wild-type mice, indicating that telomere dysfunction and accumulation of DNA damage may be of greater importance for aging of telomerase knockout mice compared to wild-type mice. Intriguingly, marker proteins of telomere dysfunction were not only detected in blood serum but also in various solid organs, including the brain, lung, kidney, liver and heart [46]. Thus, it is plausible that soluble factors that are induced by telomere dysfunction and DNA damage could be generated in many aged tissues. Interestingly, marker proteins of telomere dysfunction and DNA damage also show an age-dependent increase in human blood serum and are associated with the evolution of age-associated disease in geriatric patients [46]. These data indicate that systemic alterations induced by telomere dysfunction and DNA damage are also operative during human aging.

In summary, current data suggest alterations of the systemic environment have important roles in regulating stem cell function during aging. Aged stem cells keep the potential capacity of tissue maintenance if exposed to an optimal macroenvironment. DNA damage signaling may induce a senescence-associated secretory phenotype in the macroenvironment that contributes to stem cell aging. It is tempting to speculate that systemic factors could represent promising targets for future therapies aiming to improve stem cell function and regeneration during aging.

References

1 Castro-Malaspina H, Harris RE, Gajewski J, Ramsay N, Collins R, Dharan B, King R, Deeg HJ: Unrelated donor marrow transplantation for myelodysplastic syndromes: outcome analysis in 510 transplants facilitated by the National Marrow Donor Program. Blood 2002;99:1943–1951.

2 Kollman C, Howe CW, Anasetti C, Antin JH, Davies SM, Filipovich AH, Hegland J, Kamani N, Kernan NA, King R, Ratanatharathorn V, Weisdorf D, Confer DL: Donor characteristics as risk factors in recipients after transplantation of bone marrow from unrelated donors: the effect of donor age. Blood 2001;98:2043–2051.

3 Kantarjian H, O'Brien S, Cortes J, Giles F, Faderl S, Jabbour E, Garcia-Manero G, Wierda W, Pierce S, Shan J, Estey E: Results of intensive chemotherapy in 998 patients age 65 years or older with acute myeloid leukemia or high-risk myelodysplastic syndrome: predictive prognostic models for outcome. Cancer 2006;106:1090–1098.

4 Socie G, Stone JV, Wingard JR, Weisdorf D, Henslee-Downey PJ, Bredeson C, Cahn JY, Passweg JR, Rowlings PA, Schouten HC, Kolb HJ, Klein JP: Long-term survival and late deaths after allogeneic bone marrow transplantation. Late Effects Working Committee of the International Bone Marrow Transplant Registry. N Engl J Med 1999;341:14–21.
5 Song Z, Ju Z, Rudolph KL: Cell intrinsic and extrinsic mechanisms of stem cell aging depend on telomere status. Exp Gerontol 2009;44:75–82.
6 Nishimura EK, Granter SR, Fisher DE: Mechanisms of hair graying: incomplete melanocyte stem cell maintenance in the niche. Science 2005;307:720–724.
7 Kim JY, Siegmund KD, Tavare S, Shibata D: Age-related human small intestine methylation: evidence for stem cell niches. BMC Med 2005;3:10.
8 Ryu BY, Orwig KE, Oatley JM, Avarbock MR, Brinster RL: Effects of aging and niche microenvironment on spermatogonial stem cell self-renewal. Stem Cells 2006;24:1505–1511.
9 Shefer G, Van de Mark DP, Richardson JB, Yablonka-Reuveni Z: Satellite-cell pool size does matter: defining the myogenic potency of aging skeletal muscle. Dev Biol 2006;294:50–66.
10 Miller A, Van Zant G: Advances in hematopoietic stem cell research through mouse genetics. Curr Opin Hematol 2006;13:209–215.
11 Wagner W, Horn P, Bork S, Ho AD: Aging of hematopoietic stem cells is regulated by the stem cell niche. Exp Gerontol 2008;43:974–980.
12 Conboy IM, Conboy MJ, Wagers AJ, Girma ER, Weissman IL, Rando TA: Rejuvenation of aged progenitor cells by exposure to a young systemic environment. Nature 2005;433:760–764.
13 Haynes BF, Markert ML, Sempowski GD, Patel DD, Hale LP: The role of the thymus in immune reconstitution in aging, bone marrow transplantation, and HIV-1 infection. Ann Rev Immunol 2000;18:529–560.
14 Linton PJ, Dorshkind K: Age-related changes in lymphocyte development and function. Nat Immunol 2004;5:133–139.
15 Chambers SM, Shaw CA, Gatza C, Fisk CJ, Donehower LA, Goodell MA: Aging hematopoietic stem cells decline in function and exhibit epigenetic dysregulation. PLoS Biology 2007;5:e201:1750–1762.
16 Rossi DJ, Bryder D, Zahn JM, Ahlenius H, Sonu R, Wagers AJ, Weissman IL: Cell intrinsic alterations underlie hematopoietic stem cell aging. Proc Natl Acad Sci USA 2005;102:9194–9196.
17 Cho RH, Sieburg HB, Muller-Sieburg CE: A new mechanism for the aging of hematopoietic stem cells: aging changes the clonal composition of the stem cell compartment but not individual stem cells. Blood 2008;111:5553–5561.
18 Liang Y, Van Zant G, Szilvassy SJ: Effects of aging on the homing and engraftment of murine hematopoietic stem and progenitor cells. Blood 2005;106:1479–1487.
19 Hu X, Shen H, Tian C, Yu H, Zheng G, XuFeng R, Ju Z, Xu J, Wang J, Cheng T: Kinetics of normal hematopoietic stem and progenitor cells in a Notch1-induced leukemia model. Blood 2009;114:3783–3792.
20 Blasco MA, Lee HW, Hande MP, Samper E, Lansdorp PM, DePinho RA, Greider CW: Telomere shortening and tumor formation by mouse cells lacking telomerase RNA. Cell 1997;91:25–34.
21 Lee HW, Blasco MA, Gottlieb GJ, Horner JW 2nd, Greider CW, DePinho RA: Essential role of mouse telomerase in highly proliferative organs. Nature 1998;392:569–574.
22 Rudolph KL, Chang S, Lee HW, Blasco M, Gottlieb GJ, Greider C, DePinho RA: Longevity, stress response, and cancer in aging telomerase-deficient mice. Cell 1999;96:701–712.
23 Ju Z, Jiang H, Jaworski M, Rathinam C, Gompf A, Klein C, Trumpp A, Rudolph KL: Telomere dysfunction induces environmental alterations limiting hematopoietic stem cell function and engraftment. Nat Med 2007;13:742–747.
24 Song Z, Wang J, Guachalla LM, Terszowskif G, Rodewald HR, Ju Z, Rudolph KL: Alterations of systemic environment are the primary cause of impaired B- and T-lymphopoiesis in telomere dysfunctional mice. Blood 2009 Dec 2 [Epub ahead of print].
25 Conboy IM, Rando TA: The regulation of Notch signaling controls satellite cell activation and cell fate determination in postnatal myogenesis. Dev Cell 2002;3:397–409.
26 Conboy IM, Conboy MJ, Smythe GM, Rando TA: Notch-mediated restoration of regenerative potential to aged muscle. Science 2003;302:1575–1577.
27 Sherwood RI, Christensen JL, Conboy IM, Conboy MJ, Rando TA, Weissman IL, Wagers AJ: Isolation of adult mouse myogenic progenitors: functional heterogeneity of cells within and engrafting skeletal muscle. Cell 2004;119:543–554.
28 Cerletti M, Jurga S, Witczak CA, Hirshman MF, Shadrach JL, Goodyear LJ, Wagers AJ: Highly efficient, functional engraftment of skeletal muscle stem cells in dystrophic muscles. Cell 2008;134:37–47.
29 Shurin GV, Yurkovetsky ZR, Chatta GS, Tourkova IL, Shurin MR, Lokshin AE: Dynamic alteration of soluble serum biomarkers in healthy aging. Cytokine 2007;39:123–129.
30 Rieger MA, Hoppe PS, Smejkal BM, Eitelhuber AC, Schroeder T: Hematopoietic cytokines can instruct lineage choice. Science 2009;325:217–218.
31 Yoshihara H, Arai F, Hosokawa K, Hagiwara T, Takubo K, Nakamura Y, Gomei Y, Iwasaki H, Matsuoka S, Miyamoto K, Miyazaki H, Takahashi T, Suda T: Thrombopoietin/MPL signaling regulates hematopoietic stem cell quiescence and interaction with the osteoblastic niche. Cell stem cell 2007;1:685–697.
32 Qian H, Buza-Vidas N, Hyland CD, Jensen CT, Antonchuk J, Mansson R, Thoren LA, Ekblom M, Alexander WS, Jacobsen SE: Critical role of thrombopoietin in maintaining adult quiescent hematopoietic stem cells. Cell Stem Cell 2007;1:671–684.
33 Bersenev A, Wu C, Balcerek J, Tong W: Lnk controls mouse hematopoietic stem cell self-renewal and quiescence through direct interactions with JAK2. J Clin Invest 2008;118:2832–2844.
34 Essers MA, Offner S, Blanco-Bose WE, Waibler Z, Kalinke U, Duchosal MA, Trumpp A: IFNalpha activates dormant haematopoietic stem cells in vivo. Nature 2009;458:904–908.

35 Sato T, Onai N, Yoshihara H, Arai F, Suda T, Ohteki T: Interferon regulatory factor-2 protects quiescent hematopoietic stem cells from type I interferon-dependent exhaustion. Nat Med 2009; 15:696–700.
36 Krtolica A, Parrinello S, Lockett S, Desprez PY, Campisi J: Senescent fibroblasts promote epithelial cell growth and tumorigenesis: a link between cancer and aging. Proc Natl Acad Sci USA 2001;98:12072–12077.
37 Liu D, Hornsby PJ: Senescent human fibroblasts increase the early growth of xenograft tumors via matrix metalloproteinase secretion. Cancer Res 2007; 67:3117–3126.
38 Bavik C, Coleman I, Dean JP, Knudsen B, Plymate S, Nelson PS: The gene expression program of prostate fibroblast senescence modulates neoplastic epithelial cell proliferation through paracrine mechanisms. Cancer Res 2006; 66:794–802.
39 Coppe JP, Patil CK, Rodier F, Sun Y, Munoz DP, Goldstein J, Nelson PS, Desprez PY, Campisi J: Senescence-associated secretory phenotypes reveal cell-nonautonomous functions of oncogenic RAS and the p53 tumor suppressor. PLoS Biol 2008;6:2853–2868.
40 Ershler WB, Keller ET: Age-associated increased interleukin-6 gene expression, late-life diseases, and frailty. Ann Rev Med 2000;51:245–270.
41 Harris TB, Ferrucci L, Tracy RP, Corti MC, Wacholder S, Ettinger WH Jr, Heimovitz H, Cohen HJ, Wallace R: Associations of elevated interleukin-6 and C-reactive protein levels with mortality in the elderly. Am J Med 1999; 106:506–512.
42 Fagiolo U, Cossarizza A, Scala E, Fanales-Belasio E, Ortolani C, Cozzi E, Monti D, Franceschi C, Paganelli R: Increased cytokine production in mononuclear cells of healthy elderly people. Eur J Immunol 1993;23:2375–2378.
43 Rodier F, Coppe JP, Patil CK, Hoeijmakers WA, Munoz DP, Raza SR, Freund A, Campeau E, Davalos AR, Campisi J: Persistent DNA damage signalling triggers senescence-associated inflammatory cytokine secretion. Nat Cell Biol 2009;11:973–979.
44 Scarfi S, Fresia C, Ferraris C, Bruzzone S, Fruscione F, Usai C, Benvenuto F, Magnone M, Podesta M, Sturla L, Guida L, Albanesi E, Damonte G, Salis A, De Flora A, Zocchi E: The plant hormone abscisic acid stimulates the proliferation of human hemopoietic progenitors through the second messenger cyclic ADP-ribose. Stem Cells 2009;27: 2469–2477.
45 Schaetzlein S, Kodandaramireddy NR, Ju Z, Lechel A, Stepczynska A, Lilli DR, Clark AB, Rudolph C, Kuhnel F, Wei K, Schlegelberger B, Schirmacher P, Kunkel TA, Greenberg RA, Edelmann W, Rudolph KL: Exonuclease-1 deletion impairs DNA damage signaling and prolongs lifespan of telomere-dysfunctional mice. Cell 2007;130:863–877.
46 Jiang H, Schiffer E, Song Z, Wang J, Zurbig P, Thedieck K, Moes S, Bantel H, Saal N, Jantos J, Brecht M, Jeno P, Hall MN, Hager K, Manns MP, Hecker H, Ganser A, Dohner K, Bartke A, Meissner C, Mischak H, Ju Z, Rudolph KL: Proteins induced by telomere dysfunction and DNA damage represent biomarkers of human aging and disease. Proc Natl Acad Sci USA 2008;105: 11299–11304.

Zhenyu Ju
Max-Planck-Partner Group on Stem Cell Aging
Chinese Academy of Medical Sciences
Chaoyang District, Panjiayuan Nanli 5
Beijing 100021 (PR China)
Tel. +86 10 6777 6051, Fax +86 10 6777 6838
E-Mail zhenyuju@hotmail.com

Chapter 4

Rudolph KL (ed): Molecular Mechanisms of Adult Stem Cell Aging.
Else Kröner-Fresenius Symp. Basel, Karger, 2010, vol 1, pp 46–55

Novel Insights into Regulation of Stem Cell Aging by Aging of the Niche and Alterations in Systemic Factors

Hartmut Geiger

Department of Dermatology and Allergic Diseases, University of Ulm, Ulm, Germany

Abstract

Many tissues of the major organ systems, including the skin, intestine and hematopoietic system, are composed of short-lived cells that require continuous replenishment. This replenishment is assured by somatic stem cells. Stem cells were initially thought to be endowed with unlimited self-renewal (rejuvenation) capacity and thus assumed to be exempt from aging. This would result in a fully intact (young) stem cell compartment in chronologically aged organisms. However, evidence accumulating over the past decade has now proven that there is a measurable and successive functional decline in hematopoietic, intestinal and muscle stem cell activity from adulthood to old age, resulting in a decline of stem cell function in aged humans and mice. As stem cell activity is necessary to replenish lost differentiated cells, it has been hypothesized that aging of stem cells results in impaired tissue homeostasis in aged individuals, which might ultimately limit lifespan. In 1978, Schofield proposed the 'niche' hypothesis to describe the physiologically limited microenvironment that supports stem cells. This hypothesis has been confirmed in multiple stem cell systems from distinct organisms over the last decade, and is thus now regarded as a central theme in stem cell biology. One of the major questions in the field – and one which is an ongoing debate – is the extent of the contribution of stem cell-intrinsic versus -extrinsic (niche)/systemic factors to stem cell aging, which is important in the light of reversion/rejuvenation. While distinct aspects of aging are intrinsic to stem cells, over the recent years, the aspect of aging of the niche or even the more general influence of systemic factors on altered phenotypes associated with aged stem cells has become more and more appreciated. As this chapter demonstrates, such novel concepts (microenvironemental/systemic influence on stem cell aging) hold true for different organisms as well as different types of stem cells, rendering them a critical theme in aging of stem cell compartments. This chapter summarizes the presentations at the Else Kröner-Fresenius Symposium on the Molecular Mechanisms of Stem Cell Aging that focused primarily on elucidating the intrinsic/extrinsic contributions of aging and alterations of stem cell-niche interactions as well as systemic influences on stem cell aging. It includes presentations and novel insights into this matter in *Drosophila* and mice given by Dr. Thomas A. Rando, Dr. Amy J. Wagers, Dr. Hartmut Geiger, Dr. Ting Xie and Dr. D. Leanne Jones.

Aging of Muscle Stem Cells Is Reversible

At the Else Kröner-Fresenius Symposium on the Molecular Mechanisms of Stem Cell Aging [1–7], Dr. Thomas A. Rando presented the talk, 'Determinants and phenotypes of aged muscle stem cells'. He discussed novel aspects of his work with muscle stem cells called satellite cells (SCs), the niche and systemic factors influencing the regenerative capacity and aging of this stem cell system. He asked what are the phenotypes and molecular mechanisms of muscle stem cell aging and can they be rejuvenated?

In his previous, seminal work using heterochronic parabiotic animals (where aged and young animals are surgically tied together and thus share the same blood circulation), Dr. Rando identified that the heterochronic (young) environment was able to rejuvenate aged muscle stem cell function (see chapter 3) [8]. This meant that aging in the muscle compartment could be reverted by a systemic influence. Subsequently, lineage-tracing experiments were performed revealing that old SCs exposed to young serum behaved functionally like 'real' myogenic stem cells, while old serum converted such previously myogenic stem cells into fibrogenic cells upon activation/lineage progression. Thus, myogenic lineage tracing confirmed transdifferentiation of myogenic progenitors into a fibrogenic phenotype by old serum. The environment thus seems to play a major role in determining the fibrogenic fate of muscle stem cells. The laboratory of Dr. Rando could further demonstrate that canonical WNT signaling alters, at least in part, the fate of myogenic progenitors in response to young or old serum, as Wnt3a signaling mimicked old serum for fibrogenic conversion. Conversely, a WNT inhibitor could ameliorate the fibrotic lineage conversion induced by old serum.

Based on these results, the general question the Rando lab is currently investigating is the molecular mechanisms by which activation of the WNT pathway leads to lineage conversion. Dr. Rando shared 3 new avenues of research: (1) identification of transcriptional programs and signaling pathways by gene array analyses, (2) identification of BCL9 as a modifier of WNT signaling, and (3) identification of histone-modifying enzymes as regulators of stem cell activation.

Transcriptional Programs and Signaling Pathways

As sorting of VCAM-1$^+$CD45$^-$ cells from muscle tissue delivers a high level of pure myogenic progenitors, SCs from young and aged animals at quiescence and following activation were analyzed for gene expression profiles by array analysis. Cluster analyses revealed only small differences in general expression patterns, but activation in aged cells seemed to be delayed compared to young cells. Evolutionary conserved binding sites for WNT and FOXO signaling were used to identify WNT and FOXO target genes. Nineteen putative WNT target genes were differentially expressed in aged versus adult quiescent muscle stem cells, whereas FOXO binding sites were the only transcription factor binding sites that were actually enriched in the set of down-regulated genes, implying a possible role for FOXO signaling in transferring the WNT signaling cascade upon aging.

BCL9 as a Modifier of WNT Signaling

The proteins BCL9 and Pygopus form a quarternary complex with armadillo (or β-catenin) and TCF to initiate transcription in *Drosophila* [19]. Deletion of either BCL9 or Pygopus in *Drosophila* leads to a WNT phenotype. The role of BCL9 in mammalian systems is not very well defined, although it seems clear that it is also part of the Wnt signaling pathway. In mammals, 2 paralogs of *Drosophila* BCL9 exist: BCL9 and BCL9-2. Dr. Rando's lab analyzed the role of BCL9 and BCL9-2 in SCs. The combination of a knockdown/knockout of BCL9 and BCL9-2 in SCs postnatally resulted in a blockage of the canonical Wnt signaling pathway, while at the same time blocking the age-related lineage conversion of SCs upon activation of WNT signaling. These data might thus indicate that BCL9 and BCL9-2 are essential for the age-related phenotypic conversion of muscle cells upon WNT activation. Dr. Rando pointed out that, as published reports from Dr. Mariann Bienz's lab [20] indicate that the binary complex of BCL9 and Pygopus functions as a decoder of the histone code by binding specifically to methlylated lysine 4 on histone 3, the role of BCL9 might extend beyond that of a mere transcriptional activator.

Histone-Modifying Enzymes as Regulators of Stem Cell Activation

In a yeast 2-hybrid screen to identify transcriptional modifiers of myogenesis, the jumonji D1C gene (jmjd1c), a histone demethylase, was identified as a potential target. In combination with the BCL9 data mentioned above, this clearly makes a role for histone modification patterns in regulating young and old SCs. Microarray data then revealed that jmjd1c and another histone modifying enzyme – hairless – were highly expressed in quiescent SCs, which were down-regulated upon activation. Knocking down jmjd1c and hairless in young, but not old, quiescent SCs results in a massive increase in proliferation of SCs, indicating that young SCs are more likely influenced by epigenetic regulation in terms of histone methylation status. Subsequent analyses of global histone H3K modifications by immunohistochemistry in quiescent and activated SCs demonstrated that alterations of di- or tri-methylation on lysine 4 (which are in general activation markers) were low in quiescent cells, whereas di-or tri-methylation of lysine 9 or 27 (which in general are repression markers) was high in quiescent SCs. Further experiments are underway to get a better global picture of the methylation status of genes regulating myogenesis.

Hematopoietic Stem Cells Can Be Rejuvenated by Systemic Factors

Dr. Amy J. Wagers' interests lie in hematopoietic and skeletal muscle stem cells, their niches and their aging. These systems come with the advantage that it is feasible to prospectively identify primitive progenitor and stem cells in both the hematopoietic and muscle compartments. In her presentation at the symposium, Dr. Wagers introduced the muscle cell system by presenting data from her recent paper published in *Cell* on muscle regeneration in a dystrophic animal system [21]. In this work, the Wagers lab used cell surface marker expression coding to further purify and prospectively identify from the SC pool a distinct population of skeletal muscle precursors (SMPs, which are Sca-1$^-$, Mac-1$^-$, CD45$^-$, CXCR4$^+$, beta1-integrin$^+$) that function as muscle stem cells, based on various in vivo and transplantation assays (see chapter 7). When engrafted into muscle of dystrophin-deficient mdx mice, purified SMPs contributed to up to 94% of myofibers, restoring dystrophin expression and significantly improving muscle histology and contractile function with a regenerative index of 1 regenerated myofiber for every 28 SMPs transplanted. The fact that the transplanted SMPs also entered the PAX7$^+$ cell compartment and remained functional in secondary transplants indicated that they were able to self-renew and contribute to the endogenous stem cell pool while participating in subsequent rounds of injury repair, which could be further confirmed by BrdU incorporation assays [21].

Aging is associated with muscle atrophy. The question that was then asked in subsequent novel experiments was whether changes in the number or activity of those muscle stem cells might contribute to age-related tissue function. Using the cell surface marker code for SMPs to determine the frequency of these cells in young (2 months old) and aged mice (2 years old), it was determined that both the frequency and clongenic function (in vitro) were dramatically reduced with age. Interestingly, the decrease in SMPs with age was associated with an increase of Sca-1$^+$ cells that normally give rise to fat cells and fibroblasts in the muscle, implying that imbalances in precursor cell frequency might occur upon aging. Transcriptional profiling of SMPs from young, middle-aged and old animals revealed that middle-aged SMPs already resemble quite old SMPs in the signature profile, although their number and function are not as impaired, suggesting that alterations in transcriptional regulation are an early hallmark of aging in the muscle system. Secondly, looking at cross-tissue gene expression patterns, aged SMPs resembled more closely progenitor cells than stem cells in other tissues, suggesting that aging of SMPs results in progression along the differentiation pathway.

Hematopoietic stem cells (HSCs) from young and aged mice differ in their activity, and these differences are in part intrinsic to HSCs (see chapter 2), with additional contributions from the microenvironment (see chapter 3) [7, 18, 22–24]. We therefore refer to aged HSCs and young HSCs when we speak of HSCs from aged and young animals. What are these differences that separate young HSCs from aged HSCs? When HSCs from aged mice are transplanted along with young HSCs into a lethally irradiated recipient animal, aged HSCs are on average 2-fold less efficient in contributing to hematopoiesis compared to their young counterparts [25, 26] and exhibit a 2-fold reduced ability to home to the bone marrow (BM) upon transplantation [27]. Aged HSCs show reduced self-renewal activity as well as increased stress-induced apoptosis [28]. Aging also affects the differentiation potential of HSCs. Many studies have demonstrated that aged HSCs are deficient in their ability to support erythropoiesis, and that aged HSCs do not efficiently generate T and B lymphoid progeny, while they are superior at supporting the myeloid cell lineage (see [29] and references cited therein). Aged HSCs have also been reported to show increased levels of gamma H2AX staining, a likely surrogate marker for DNA double-strand breaks [30].

HSCs reside in specialized 3-dimensional structures in the BM that are referred to as niches. In BM, 2 types of niches have been identified: the vascular niche and the endosteal niche. Several reports support the notion that HSCs in BM are closely located to the endosteum of the bone [31–39]. In the endosteal niche, HSCs are located in close proximity to stroma cells and extra cellular matrix components forming the niche. The endosteal bone surface is primarily populated by osteoblasts. It is therefore believed that the stroma in the niche mainly consists of osteoblasts and that the reciprocal communication with and adhesion to osteoblasts plays a central role in regulating self-renewal and differentiation of HSCs [39–43]. Osteoblasts are known to express a variety of adhesion molecules and to produce a variety of cytokines acting on HSCs [41–43]. The interaction of HSCs with the niche and thus with stromal cells/osteoblasts is regarded as central to the biology of hematopoiesis [40, 44, 45]. HSCs in BM are also found in the vicinity of sinusoid endothelial cells, referred to as the vascular niche [33]. The existence of niches distinct from the endosteal niche is further supported by the fact that the spleen, an organ that lacks osteoblasts and bone, harbors HSCs. It is not clear at the moment whether the niche in the spleen is of endothelial type [33]. It has been suggested that this vascular niche forms a milieu that supports proliferation, differentiation and trans-endothelial migration of HSCs, whereas the endosteal niche supports quiescence and self-renewal. New reports support the spatial coexistence of both niches, especially in the trabecular part of the bone, possibly rendering the distinction between endosteal and vascular more semantic [46].

In the field of HSC aging, there is an ongoing debate regarding the extent to which aging is stem cell intrinsic and what contribution the niche and aged systemic environment may make. Building on her established experience with heterochronic parabiosis, the effect of heterochronic parabiosis on aging of HSCs was determined (see chapter 3).

In heterochronic parabiosis, aged and young animals are surgically tied together and thus share the same blood circulation. In Dr. Wagers' laboratory, blood chimerism reaches 50% by

7–10 days, with a rapid exchange of cells and factors across the vascular junction. Animals are usually analyzed 4–5 weeks after initiation of parabiosis. As mentioned above, upon aging the number of phenotypically defined stem cells increases in C57BL/6 animals. As a first sign of rejuvenation, the number of HSCs in aged animals was almost reverted to the frequency found in young animals upon exposure to young blood in the heterochronic parabiosis. More interestingly, heterochronic parabiosis corrected the engraftment dysfunction associated with aged HSCs, meaning that stem cells isolated from aged heterochronically parabiosed animals were significantly better in reconstitution of recipients compared to cells isolated from aged isochronic control animals. Additional experiments revealed that the rejuvenation of HSCs was mediated by changes in osteolineage stromal support cells (niche cells). The number of osteolineage stroma cells, similar to HSCs, expanded in C57BL/6 animals with age, which was corrected/reverted by exposure to young blood in the heterochronic situation. More importantly, the functional decline of these cells was ameliorated. It was also possible to obtain functionally aged stem cells by exposing young stem cells to osteolineage stroma cells isolated from aged animals. Aged osteolineage cells could not exert that influence anymore after they were exposed systemically to young blood via parabiosis. Osteolineage cells from aged animals under heterochronic parabiosis also reverted the lineage differentiation bias of aged HSCs (myeloid over lymphoid), thus rejuvenating the function of aged HSCs on multiple levels.

Dr. Wagers concluded her presentation by emphasizing that the changes in stem cell function as a consequence of heterochronic parabiosis obviously happen in multiple tissues, with the exciting news that rejuvenation of aged stem cell function is possible.

Adhesion Is Central to Stem Cell Niche Interactions

Which mechanisms retain HSCs in their position in the niche and ensure a stable interaction with stroma cells? The repertoire, level of expression and functional status of adhesion molecules expressed on HSCs as well as on stroma cells forming the niche determine the ability of the HSC to adhere to its niche. HSCs adhere to the niche by cell-cell contacts as well as cell-extracellular matrix contacts. The precise molecular signals that are exchanged between cells forming the niche and HSCs still remain largely elusive, which is most likely due to the limited experimental accessibility of the niche in vivo. Interestingly, a distinct role for altered expression of adhesion molecules in HSC differentiation was recently suggested by Forsberg et al. [47], supporting a model in which altered adhesion of aged HSCs to stroma cells in the niche will influence the biology of aged HSCs. Adhesion and detachment of HSCs require significant changes in the actin cytoskeleton as well as the activity of adhesion receptors. These intracellular changes are in part regulated by outside-in signaling of receptors sensing the cellular environment, which then result in inside-out reactions of the cell and either adhesion or detachment. This outside-in information is transmitted through members of the family of small Rho GTPases and other signaling pathways [48–50]. Small Rho GTPases, including Rac1, Rac2, Cdc42 and RhoA, are primarily involved in the regulation of adhesion signaling in response to outside signals in various mammalian cell types, including hematopoietic cells, and the individual contribution of these small Rho GTPases to regulation is cell type specific as well as adhesion substrate specific [31, 51–55].

Dr. Hartmut Geiger's laboratory previously published work indicating that primitive hematopoietic cells (hematopoietic progenitor cells, HPCs) in aged mice show reduced adhesion to

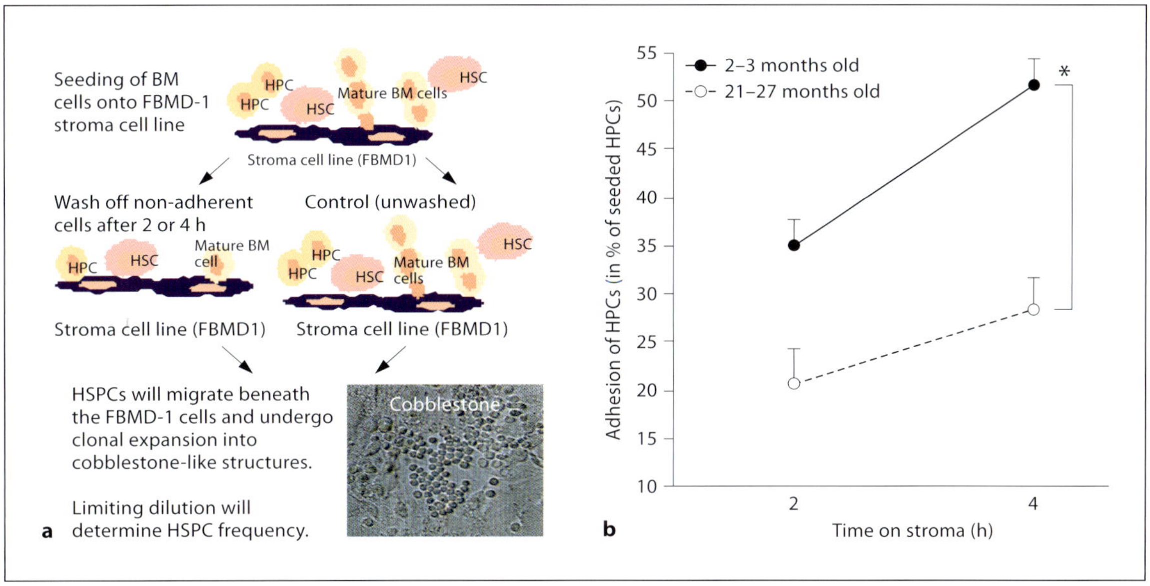

Fig. 1. Hematopoietic progenitor cells (HPCs) from aged mice are reduced in their ability to adhere to stroma. **a** BM cells from aged (21–27 months) and young mice (2–3 months) were subjected to a CAFC adhesion assay to determine adhesion of hematopoietic stem and progenitor cells (HSPCs). Experimental setup. **b** Percentage of adherent HPCs after 2 or 4 h determined as CAFC day-7 cells using the CAFC adhesion assay. * $p < 0.05$.

stroma cells (fig. 1), which correlated with elevated levels of activity of the small Rho GTPase Cdc42 compared to cells from young mice, but no change in the activity of Rac1 and Rac2 [56]. This led to the hypothesis that, in contrast to their young counterparts, aged primitive hematopoietic cells present altered, less favorable interactions with the niche, which might in part explain impaired hematopoiesis in aged individuals. In collaboration with Dr. Gunzer in Magdeburg (Germany), the laboratory of Dr. Geiger thus developed time-lapse multiphoton intravital microscopy (MP-IVM) in long bones to determine the dynamics of young and aged hematopoietic cells inside the diaphysis of a long bone inside the marrow cavity. Besides introducing a novel approach for intravital imaging of individual cells within long bones, they determined that differentiated hematopoietic cells, such as macrophages and dendritic cells, residing in BM are motile while showing a distinct localization with respect to the endosteal surface. This was in contrast to the finding that young early HPCs (LIN^-, $Sca\text{-}1^+$, $C\text{-}Kit^+$ or eHPCs), although remaining fixed on the spot, presented with constantly ongoing cell protrusion movements indicating synapse-like interactions with stroma cells in the diaphyseal regions of long bones in vivo. This was determined with a new mathematical algorithm to quantify cell protrusion movement from 3-dimensional time-lapse imaging data. As anticipated, HPCs and eHPCs were found in distinct locations relative to the endosteum. Finally, the data presented demonstrated that aged eHPCs display a higher cell protrusion activity and are localized more distantly from the endosteum compared to young eHPCs, while showing reduced adhesion to stroma cells and presenting with a reduced number of polarized cells upon adhesion.

In the aggregate, the results presented may provide insights into the dynamics and physiology of aged primitive cells in the marrow of long bones in vivo. This might form a rational basis to explain the reduced hematopoietic performance of aged individuals. Dr. Geiger concluded that elevated activity of Cdc42 might play a critical role in adult hematopoiesis in the BM and that elevated activity of Cdc42 might be causative for changes in the function of aged HSCs via altered interaction with the niche.

Drosophila Germ Line Stem Cells and Their Niche

The *Drosophila* ovary contains at least 3 types of stem cells: germ line stem cells (GSCs), escort stem cells, and follicular stem cells, which work together to efficiently assemble egg chambers. Of these 3 types of stem cells, the GSC was the first to be shown to be controlled by the niche, due to its easy identification and well-defined surrounding cells. There are 2 or 3 GSCs per germarium and up to 5–7 niche cells. The laboratory of Dr. Ting Xie has shown in the past that the niche (called cap cells) controls GSC self-renewal, anchorage, aging and competition, and the GSC also signals back to the niche for its maintenance. In addition, it turned out that follicular stem cells are an attractive model for studying epithelial stem cell regulation and signal integration as they resemble mammalian epithelial stem cells and require multiple signaling pathways for their self-renewal [16]. *Drosophila* BMPs (JBP, DPP) expressed by the niche cells act directly on GSC to prevent differentiation, while deletion of adhesion of GSCs by deletion of cadherin results in the loss of GSCs [57, 58]. In addition, notch signaling in GSCs is critical for their maintenance. Sophisticated tools allow the manipulation of gene function of ovary GSCs and their niche: the gal4-UAS system can be used to drive niche or GSC-specific gene over-expression or RNAi knockdowns, while flip recombinase-mediated techniques can be applied to inactivate gene function specifically in stem cells and their progeny.

In his presentation 'Genetic control of germline stem cell aging in the *Drosophila* ovary', Dr. Xie addressed 2 basic questions in stem cell aging biology: (1) how do we define stem cell aging in the *Drosophila* ovary? (2) How is it mechanistically controlled? As reported for many stem cell systems, the number of *Drosophila* ovary stem cells, their proliferation rate and the number of niche cells all decline with age (from a 7-day-old fly to a 60-day-old fly). The laboratory of Dr. Xie identified that BMP signaling activity that originates from the niche declines with age. By reducing the BMP dosage by genetic means to mimic an aged niche, the age-dependent loss of GSC was further enhanced. Stem cell-intrinsic as well as niche-specific increased expression of BMP signaling activity using the gal4-UAS system slowed down the age-dependent loss of GSCs and enhanced the proliferation rate of GSCs. With respect to adhesion, E-cadherin expression in the stem cell-niche junction declined with age. Reducing the dosage of E-cadherin further enhanced stem cell aging, while increasing E-cadherin dosage slowed down the loss of stem cells with age. These findings strongly support that reduced niche signaling and stem cell niche adhesion contribute to the age-dependent decline in stem cell number and proliferation.

The *Drosophila* adult male testis is a long, coiled tube filled with cells at all stages of spermatogenesis. In adult male flies, 9 GSCs lie at the apical tip of the testis, forming a ring that closely surrounds a cluster of post-mitotic somatic cells, called the hub. Upon assymetrical division, a male GSC normally gives rise to 1 cell that will retain stem cell identity and 1 cell, called a gonialblast, that is displaced away from the hub and initiates differentiation [59]. The gonialblast and its progeny will then undergo 4 rounds of mitotic divisions, with incomplete cytokinesis, creating a cluster of 16 interconnected spermatogonia. The JAK/STAT pathway is required for self-renewal of the male GSCs [59]. The ligand that activates the *Drosophila* JAK/STAT pathway, the signaling protein unpaired (Upd), is normally expressed in the 'apical hub' cells. The Upd ligand is glycosylated and is tightly associated with the extracellular matrix after secretion [60]. Consequently, hub cells have been proposed to contribute to the male GSC niche, instructing the adjacent germ cells to maintain stem cell identity following division by providing a local source of the critical signal Upd. Similar to the hematopoietic system in mammals and the female germ line in *Drosophila*, cell adhesion between niche cells and stem cells has also been shown to be required for stem cell maintenance in the male germ line in *Drosophila*. Clusters of adherens junctions are present between male GSCs and the adjacent hub cells [59]. It is believed that cell adhesion is required for holding stem cells within the niche in close proximity to maintenance signals and away from differentiation cues [61].

The laboratory of Dr. D. Leanne Jones uses the *Drosophila* male germ line system as a model to study the role of the stem cell niche for stem cell maintenance. As is the case for stem cells in the female ovary, the key components in the male testis can be well visualized. Germ cells in the testis can be labeled by an antibody against the Vasa protein, with germ line stem cells being in close contact with the post-somatic hub, the niche, which is easily visualized by specific staining for hub cells. There is decreased spermatogenesis in the aged male *Drosophila* germ line that correlates with a reduced stem cell number and also a reduced number of hub cells. The laboratory of Dr. Jones has reported in the past how the process of aging affects GSC behavior in the *Drosophila* testis. Investigating closely the role of the microenvironment in that process, they reported that significant changes within the stem cell niche (the hub) occurred that contributed to a decline in stem cell number over time. Hub cells in testes from older males, similar to niche cells from aged ovaries in *Drosophila*, displayed reduced expression of the cell adhesion molecule cadherin and the key self-renewal signal protein Upd, which activates the JAK/STAT pathway. Interestingly, forced expression of Upd within niche cells maintained GSCs in older males and thus ameliorated stem cell aging [62].

Given the importance of hub cells for stem cells, the Jones lab used a number of novel experiments to look at where hub cells come from and how they are regulated. Lineage-tracing experiments with a heat shock-inducible system for recombination to induce lacZ as a lineage marker revealed that somatic stem cells (SSCs) contributed to the apical hub (with a peak at 10 days after induction) and thus these SSCs are bi-potent, as they are also able to differentiate into so-called cis-cells. To determine the factors that regulate this transition, 2 candidate genes were further analyzed: E-cadherin and escargot. E-cadherin turned out not to be required for SSCs to contribute to the apical hub. However, the transcription factor escargot, when deleted in SSCs, was cell-autonomously required to maintain the integrity of the apical hub, and thus differentiation of SSCs from hub cells. Finally, Dr. Jones presented data on a screen using a temperature-sensitive Gal4-UAS system for temporal and spatial expression of an RNAi library to identify factors cell-autonomously required for hub maintenance. STAT92E was identified as a target gene, as it was cell-autonomously necessary in hub cells for the maintenance of hub and stem cells. Another hit, CG7646, most likely a transcription factor, was identified to be necessary for hub cell maintenance and, interestingly, hub cells rebuild upon reactivation of CG7646. In aggregation, the data presented demonstrate that molecular pathways regulating hub/niche are potent regulators of stem cells and are themselves regulated by multiple pathways.

References

1 Morrison SJ, Uchida N, Weissman IL: The biology of hematopoietic stem cells. Ann Rev Cell Dev Biol 1995;11: 35–71.
2 Tani H, Morris RJ, Kaur P: Enrichment for murine keratinocyte stem cells based on cell surface phenotype. Proc Natl Acad Sci USA 2000;97:10960–10965.
3 Fuchs E, Segre JA: Stem cells: a new lease on life. Cell 2000;100:143–155.
4 Potten CS, Morris RJ: Epithelial stem cells in vivo. J Cell Sci Suppl 1988;10: 45–62.
5 Stappenbeck TS, Mills JC, Gordon JI: Molecular features of adult mouse small intestinal epithelial progenitors. Proc Natl Acad Sci USA 2003;100: 1004–1009.
6 Van Zant G, Liang Y: The role of stem cells in aging. Exp Hematol 2003;31: 659–672.
7 Geiger H, Van Zant G: The aging of lympho-hematopoietic stem cells. Nat Immunol 2002;3:329–333.
8 Conboy IM, Conboy MJ, Wagers AJ, Girma ER, Weissman IL, Rando TA: Rejuvenation of aged progenitor cells by exposure to a young systemic environment. Nature 2005;433:760–764.
9 Kirkwood TB: Intrinsic ageing of gut epithelial stem cells. Mech Ageing Dev 2004;125:911–915.
10 Potten CS, Martin K, Kirkwood TB: Ageing of murine small intestinal stem cells. Novartis Found Symp. 2001;235: 66–79; discussion 79–84, 101–104.
11 Kim M, Moon HB, Spangrude GJ: Major age-related changes of mouse hematopoietic stem/progenitor cells. Ann NY Acad Sci 2003;996:195–208.
12 Sudo K, Ema H, Morita Y, Nakauchi H: Age-associated characteristics of murine hematopoietic stem cells. J Exp Med 2000;192:1273–1280.
13 Geiger H, True JM, De Haan G, Van Zant G: Longevity and stem cells: a genetic connection. Sci World J 2001;1:77.
14 Sharpless NE, DePinho RA: Telomeres, stem cells, senescence, and cancer. J Clin Invest 2004;113:160–168.
15 Torella D, Rota M, Nurzynska D, et al: Cardiac stem cell and myocyte aging, heart failure, and insulin-like growth factor-1 overexpression. Circ Res 2004; 94:514–524.
16 Li L, Xie T: Stem cell niche: structure and function. Annu Rev Cell Dev Biol 2005;21:605–631.
17 Papayannopoulou T, Scadden DT: Stem-cell ecology and stem cells in motion. Blood 2008;111:3923–3930.
18 Ju Z, Jiang H, Jaworski M, Rathinam C, Gompf A, Klein C, Trumpp A, Rudolph KL: Telomere dysfunction induces environmental alterations limiting hematopoietic stem cell function and engraftment. Nat Med 2007;13:742–747.
19 Mieszczanek J, de la Roche M, Bienz M: A role of Pygopus as an anti-repressor in facilitating Wnt-dependent transcription. Proc Natl Acad Sci USA 2008; 105:19324–19329.
20 Fiedler M, Sanchez-Barrena MJ, Nekrasov M, Mieszczanek J, Rybin V, Muller J, Evans P, Bienz M: Decoding of methylated histone H3 tail by the Pygo-BCL9 Wnt signaling complex. Mol Cell 2008;30:507–518.
21 Cerletti M, Jurga S, Witczak CA, Hirshman MF, Shadrach JL, Goodyear LJ, Wagers AJ: Highly efficient, functional engraftment of skeletal muscle stem cells in dystrophic muscles. Cell 2008; 134:37–47.
22 Geiger H, Rennebeck G, Van Zant G: Regulation of hematopoietic stem cell aging in vivo by a distinct genetic element. Proc Natl Acad Sci USA 2005; 102:5102–5107.
23 Rossi DJ, Bryder D, Zahn JM, Ahlenius H, Sonu R, Wagers AJ, Weissman IL: Cell intrinsic alterations underlie hematopoietic stem cell aging. Proc Natl Acad Sci USA 2005;102:9194–9199.
24 Rando TA: Stem cells, ageing and the quest for immortality. Nature 2006; 441:1080–1086.
25 Morrison SJ, Wandycz AM, Akashi K, Globerson A, Weissman IL: The aging of hematopoietic stem cells. Nat Med. 1996;2:1011–1016.
26 Chen J, Astle CM, Harrison DE: Genetic regulation of primitive hematopoietic stem cell senescence. Exp Hematol 2000;28:442–450.
27 Liang Y, Van Zant G, Szilvassy SJ: Effects of aging on the homing and engraftment of murine hematopoietic stem and progenitor cells. Blood 2005; 106:1479–1487.
28 Janzen V, Forkert R, Fleming HE, Saito Y, Waring MT, Dombkowski DM, Cheng T, DePinho RA, Sharpless NE, Scadden DT: Stem-cell ageing modified by the cyclin-dependent kinase inhibitor p16INK4a. Nature 2006;443:421–426.
29 Linton PJ, Dorshkind K: Age-related changes in lymphocyte development and function. Nat Immunol 2004;5: 133–139.
30 Rossi DJ, Bryder D, Seita J, Nussenzweig A, Hoeijmakers J, Weissman IL: Deficiencies in DNA damage repair limit the function of haematopoietic stem cells with age. Nature 2007;447: 725–729.
31 Cancelas JA, Lee AW, Prabhakar R, Stringer KF, Zheng Y, Williams DA: Rac GTPases differentially integrate signals regulating hematopoietic stem cell localization. Nat Med 2005;11:886–891.
32 Nilsson SK, Johnston HM, Coverdale JA: Spatial localization of transplanted hemopoietic stem cells: inferences for the localization of stem cell niches. Blood 2001;97:2293–2299.
33 Kiel MJ, Yilmaz OH, Iwashita T, Terhorst C, Morrison SJ: SLAM family receptors distinguish hematopoietic stem and progenitor cells and reveal endothelial niches for stem cells. Cell 2005; 121:1109–1121.
34 Nilsson SK, Simmons PJ: Transplantable stem cells: home to specific niches. Curr Opin Hematol 2004;11:102–106.
35 Adams GB, Chabner KT, Alley IR, Olson DP, Szczepiorkowski ZM, Poznansky MC, Kos CH, Pollak MR, Brown EM, Scadden DT: Stem cell engraftment at the endosteal niche is specified by the calcium-sensing receptor. Nature 2005;439:599–603.
36 Calvi LM, Adams GB, Weibrecht KW, Weber JM, Olson DP, Knight MC, Martin RP, Schipani E, Divieti P, Bringhurst FR, Milner LA, Kronenberg HM, Scadden DT: Osteoblastic cells regulate the haematopoietic stem cell niche. Nature 2003;425:841–846.
37 Katayama Y, Battista M, Kao WM, Hidalgo A, Peired AJ, Thomas SA, Frenette PS: Signals from the sympathetic nervous system regulate hematopoietic stem cell egress from bone marrow. Cell 2006;124:407–421.

38 Cancelas JA, Williams DA: Stem cell mobilization by beta2-agonists. Nat Med 2006;12:278–279.
39 Zhang J, Niu C, Ye L, Huang H, He X, Tong WG, Ross J, Haug J, Johnson T, Feng JQ, Harris S, Wiedemann LM, Mishina Y, Li L: Identification of the haematopoietic stem cell niche and control of the niche size. Nature 2003; 425:836–841.
40 Yin T, Li L: The stem cell niches in bone. J Clin Invest 2006;116:1195–1201.
41 Jung Y, Wang J, Havens A, Sun Y, Jin T, Taichman RS: Cell-to-cell contact is critical for the survival of hematopoietic progenitor cells on osteoblasts. Cytokine 2005;32:155–162.
42 Neiva K, Sun YX, Taichman RS: The role of osteoblasts in regulating hematopoietic stem cell activity and tumor metastasis. Braz J Med Biol Res 2005; 38:1449–1454.
43 Taichman RS: Blood and bone: two tissues whose fates are intertwined to create the hematopoietic stem-cell niche. Blood 2005;105:2631–2639.
44 Adams GB, Scadden DT: The hematopoietic stem cell in its place. Nat Immunol 2006;7:333–337.
45 Scadden DT: The stem-cell niche as an entity of action. Nature 2006;441:1075–1079.
46 Lo Celso C, Fleming HE, Wu JW, Zhao CX, Miake-Lye S, Fujisaki J, Cote D, Rowe DW, Lin CP, Scadden DT: Live-animal tracking of individual haematopoietic stem/progenitor cells in their niche. Nature 2009;457:92–96.
47 Forsberg EC, Prohaska SS, Katzman S, Heffner GC, Stuart JM, Weissman IL: Differential expression of novel potential regulators in hematopoietic stem cells. PLoS Genet 2005;1:e28.
48 Hall A: G proteins and small GTPases: distant relatives keep in touch. Science 1998;280:2074–2075.
49 Hall A: Rho GTPases and the actin cytoskeleton. Science 1998;279:509–514.
50 Defilippi P, Valles AM: The winding road from adhesive receptors to the nucleus. Conference on molecular biology of cellular interactions: adhesion receptor signalling and regulation of gene expression. EMBO Rep 2002;3: 312–316.
51 Yang FC, Atkinson SJ, Gu Y, Borneo JB, Zheng Y, Pennington J, Williams DA: Rac and Cdc42 GTPases control hematopoietic stem cell shape, adhesion, migration, and mobilization. Proc Natl Acad Sci USA 2001;98:5614–5618.
52 Gu Y, Filippi MD, Cancelas JA, Siefring JE, Williams EP, Jasti AC, Harris CE, Lee AW, Prabhakar R, Atkinson SJ, Kwiatkowski DJ, Williams DA: Hematopoietic cell regulation by Rac1 and Rac2 guanosine triphosphatases. Science 2003;302:445–449.
53 Jansen M, Yang FC, Cancelas JA, Bailey JR, Williams DA: Rac2-deficient hematopoietic stem cells show defective interaction with the hematopoietic microenvironment and long-term engraftment failure. Stem Cells 2005; 23:335–346.
54 Cancelas JA, Jansen M, Williams DA: The role of chemokine activation of Rac GTPases in hematopoietic stem cell marrow homing, retention, and peripheral mobilization. Exp Hematol 2006; 34:976–985.
55 Ghiaur G, Lee A, Bailey J, Cancelas J, Zheng Y, Williams DA: Inhibition of RhoA GTPase activity enhances hematopoietic stem and progenitor cell proliferation and engraftment in vivo. Blood 2006;108:2087–2094.
56 Xing Z, Ryan MA, Daria D, Nattamai KJ, Van Zant G, Wang L, Zheng Y, Geiger H: Increased hematopoietic stem cell mobilization in aged mice. Blood 2006;108:2190–2197.
57 Song X, Xie T: DE-cadherin-mediated cell adhesion is essential for maintaining somatic stem cells in the *Drosophila* ovary. Proc Natl Acad Sci USA 2002; 99:14813–14818.
58 Song X, Zhu CH, Doan C, Xie T: Germline stem cells anchored by adherens junctions in the *Drosophila* ovary niches. Science 2002;296:1855–1857.
59 Yamashita YM, Fuller MT, Jones DL: Signaling in stem cell niches: lessons from the *Drosophila* germline. J Cell Sci 2005;118:665–672.
60 Harrison DA, McCoon PE, Binari R, Gilman M, Perrimon N: *Drosophila* unpaired encodes a secreted protein that activates the JAK signaling pathway. Genes Dev 1998;12:3252–3263.
61 Gonzalez-Reyes A: Stem cells, niches and cadherins: a view from *Drosophila*. J Cell Sci 2003;116:949–954.
62 Boyle M, Wong C, Rocha M, Jones DL: Decline in self-renewal factors contributes to aging of the stem cell niche in the *Drosophila* testis. Cell Stem Cell 2007;1:470–478.

Hartmut Geiger
Klinik für Dermatologie und Allergologie, Universität Ulm
James-Franck-Ring 11
DE–89081 Ulm (Germany)
Tel. +49 731 5005 7650, Fax +49 731 5005 7652
E-Mail hartmut.geiger@uni-ulm.de

Rudolph KL (ed): Molecular Mechanisms of Adult Stem Cell Aging.
Else Kröner-Fresenius Symp. Basel, Karger, 2010, vol 1, pp 56–62

Aging and Hematopoietic Malignancies

Anne Gompf

Institute of Molecular Medicine and Max Planck Research Group on Stem Cell Aging, University of Ulm, Ulm, Germany

Abstract

This chapter summarizes current data indicating how aging of hematopoietic stem cells and the stem cell environment can contribute to leukemia development and progression. Cancer arises through the progressive genetic and epigenetic alterations of normal cells, and according to the current view, 5–7 hits are required for the transformation of a normal cell into a cancer cell. These events enable transformed cells to gain immortal proliferation, the ability to evade recognition by the immune system, and to evade cell cycle control and apoptosis induction. One important concept in cancer formation indicates that the cancer-initiating cell type includes stem cells, since stem cells naturally possess features that are also important for tumor growth: the potential for self-renewal, increased proliferative potential, and the potential to generate (differentiated) progeny. Alternatively, tumor cells may arise from more committed, non-stem cells that acquire stem-like features during the course of transformation. This report summarizes current findings that were presented at the Else Kröner-Fresenius Symposium on the Molecular Mechanisms of Adult Stem Cell Aging indicating that aging induces alterations at stem and progenitor cell level influencing the development of hematopoietic malignancies.

Loss of Proliferative Capacity of Hematopoietic Cells Can Accelerate Leukemogenesis

A series of reviews has summarized molecular events that underlie the transformation of hematopoietic cells into cancer cells [1, 2] (summarized in fig. 1).

Hematological malignancies include (1) leukemias (Greek for white blood), which are cancers of the circulating blood and the bone marrow, (2) lymphomas, which are primarily solid tumors of the lymph nodes, and (3) related disorders such as myelodysplastic syndromes. According to the differentiation state, leukemia can be divided into acute and chronic leukemia consisting of rather immature or mature leukocytes, respectively. Leukemias are phenotypically classified as leukemias of myeloid origin [acute myeloid leukemia (AML) and chronic myeloid leukemia (CML)], or leukemias of lymphoid origin [acute lymphoid leukemia (ALL) and chronic lymphoid leukemia].

A well-studied example of adult leukemia is CML, which is consistently associated with a genetic abnormality called the Philadelphia chromosome – the consequence of a reciprocal trans-

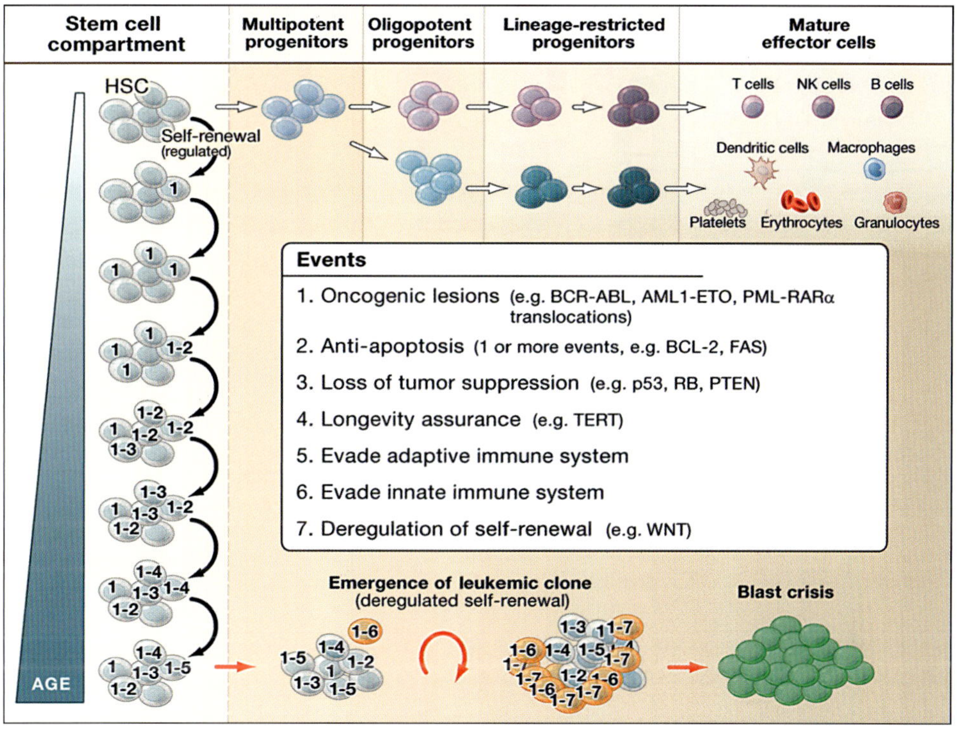

Fig. 1. Model of leukemic progression. As HSCs possess the ability for self-renewal, they accumulate genetic and epigenetic lesions throughout the lifetime, leading to oncogene activation, apoptosis inhibition, loss of tumor suppressors, activation of longevity assurance genes (telomerase), evasion from immune responses, and increased self-renewal. The initial lesion is indicated as '1' in the cells on the left side of the figure. The accumulation of these lesions – indicated as '1–2' or '1–3' and so on – can lead to leukemia formation. The actual number of lesions that have to be acquired before a cell transforms into a leukemic cell may vary as 1 lesion can confer various malignant properties. It has been shown that some of the initial events occur at the stem cell level, whereas some of the later events can also occur at progenitor cell level, leading to a reactivation of stem cell properties (e.g. self-renewal) and to malignant transformation. Reproduced with permission from Rossi et al. [2].

location between human chromosomes 9 and 22 [3, 4]. The resulting BCR-ABL fusion oncogene codes for a constitutively active tyrosine kinase responsible for transmitting mitogenic and anti-apoptotic signals [5]. Besides CML, ALL can also result from the rearrangement of the BCR and ABL genes. In CML, the translocation has been shown to occur on the level of hematopoietic stem cells (HSCs) or cells that acquired self-renewal potential [6, 7], whereas in ALL the translocation is thought to occur in lymphoid-committed progenitor cells (recently reviewed in [8]).

Hence, BCR-ABL has been a model to investigate the mechanisms of leukemia development for many years. It has been shown that the introduction of BCR-ABL fusion genes into bone marrow progenitor cells can induce B-ALL (B cell acute lymphoblastic leukemia; upon introduction of the p190 variant of BCR-ABL) or CML (upon introduction of p210 BCR-ABL) in mice, closely resembling the human diseases caused by the respective variants of the fusion protein [9–11].

In humans, aging of the hematopoietic system is associated with an impairment in hemato-lymphopoiesis resulting in reduced lymphopoiesis and anemia. On the contrary, myelopoiesis seems to be increased in the elderly, although erythrocytes as a product of myelopoiesis are decreased as well. The exact molecular basis of these changes is still under debate and may include DNA damage-induced increases in senescence and apoptosis (see chapter 9), changes in the stem cell environment induced by telomere dysfunction and aging (see chapters 3 and 4), or alterations in gene expression and epigenetic control leading to stem cell dysfunction (see chapter 2). An emerging research area is to analyze the effects of these complex mechanisms of HSC aging on the development of hematopoietic malignancies.

In 2005, Prof. James DeGregori and colleagues showed in transplantation-based mouse experiments that the selection of BCR-ABL-transduced hematopoietic cells upon transplantation in mice and the evolution of full-blown leukemia are strongly dependent on the proliferative competition of the non-transformed competitor bone marrow. Specifically, the study showed that proliferation-defective bone marrow cells cannot compete against BCR-ABL-transduced hematopoietic cells, and even a small fraction of transplanted BCR-ABL-positive bone marrow progenitor cells leads to rapid development of leukemia. In contrast, proliferation-competent bone marrow cells can successfully compete against BCR-ABL-positive bone marrow cells, thus preventing the evolution of leukemia [11].

New data from the DeGregori laboratory were presented at the Else Kröner-Fresenius Symposium on the Molecular Mechanisms of Adult Stem Cell Aging showing that low-dose γ-irradiated bone marrow cells also possess reduced competitive potential, resulting in enhanced selection of oncogene-positive cells [12]. It remains to be analyzed whether DNA damage contributes to the enhanced selection of leukemic cells and increased incidence of leukemia during aging. It is conceivable that the age-related accumulation of DNA damage [13] may have similar effects on decreasing the proliferative competition of bone marrow cells as seen in response to low-dose γ-irradiation in mice.

Prof. DeGregori presented recent data from his group supporting the concept that a loss of proliferative competition occurs with natural aging of hematopoietic cells in wild-type mice and that this defect can contribute to the selection of malignant clones and leukemogenesis. Lethally irradiated recipient mice were transplanted with low-percentage BCR-ABL-positive bone marrow cells mixed with unfractionated healthy competitor bone marrow cells from young or old mice. Competitor cells deriving from young mice prevented leukemia development. In contrast, half of the mice reconstituted with BCR-ABL-positive progenitor cells mixed with competitor cells from old mice developed leukemia within 75 days after transplantation.

Prof. DeGregori argues from an evolutionary point of view that cellular fitness declines with age, leading to impaired proliferation capacity. According to his hypothesis, stem cells from young individuals are 'fit' and well-adapted to their tissue environment, and an oncogenic mutation cannot confer better adaptation to the target cells (fig. 2). Therefore, these mutant cells lose the competition to repopulate bone marrow niches against the well-adapted young cell selected during natural evolution. In contrast, genetic and epigenetic changes and alterations in the environment occur during aging, limiting the fitness of HSCs (see chapters 2–4). In this aged cellular context, the same oncogenic mutations that are non-adaptive in a young stem cell pool can now show an adaptive advantage over old competitors in terms of proliferation or apoptosis evasion (fig. 2). According to this hypothesis, the aging phenotype is tumor-promoting by selecting adaptive oncogenic mutations, whereas the young environment is inherently tumor-suppressive. However, Prof. DeGregori did not directly analyze homing

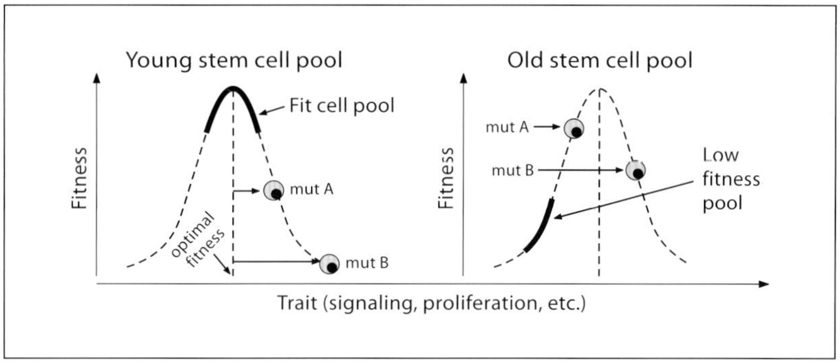

Fig. 2. Adaptive oncogenesis model of leukemia development and selection for mutations in young and old environments. In young organs, the non-transformed young cells have an optimal level of cellular fitness and most oncogenic mutations will result in decreased cellular fitness, resulting in impaired competitive selection compared to non-transformed cells. However, in old organs, aged cells exhibit reduced fitness and the same oncogenic mutations can now result in improved fitness, conferring a selective advantage to the mutant cells compared to non-transformed, aged cells. Reproduced with permission from Marusyk and DeGregori [14].

and engraftment properties for both young and old cells in his experimental setting of competitive transplantation. Therefore, it is not clear whether leukemic stem cells would really compete for the same niche as the normal HSCs. In addition, it will be important to analyze whether the hypothesis of age-dependent changes in proliferative competition of stem and progenitor cells translates to other organ systems and to human aging.

The meeting also showed that the influence of aging on hematopoietic malignancies is complex. Prof. Kenneth Dorshkind and colleagues [15] provided evidence in studies on leukemia formation that cell-intrinsic effects can influence tumorigenesis. New work from his group was presented at the meeting, providing the first experimental evidence that age-dependent cell-intrinsic changes increase the resistance of B lymphocyte progenitor cells to malignant transformation [16]. According to the data of Prof. Dorshkind and colleagues on the transformation resistance of aged B lymphocyte progenitor cells, it can be postulated that both declining tissue homeostasis and decreased transformation ability of old cells are consequences of the same molecular mechanisms (senescence and apoptosis). In addition to reduced proliferation, aged cells may stimulate immune responses that lead to a clearance of aged and damaged cells from tissues. It is conceivable that oncogenic mutations would stimulate such immune regulatory mechanisms in aged cells but to a lesser extent in young cells, resulting in an improved immune clearance of mutant cells during aging.

In summary, it is possible that both mechanisms, an age-dependent increase in cellular resistance to transformation and a decreasing fitness of competing progenitor cells, influence the evolution of malignancies during aging. The relative contribution of these different mechanisms to malignancies that occur during human aging remains to be investigated. In addition, it would be interesting to analyze whether age-dependent changes in the systemic environment can contribute to the development of leukemia.

Leukemic Cells Evade the Immune System

In addition to reduced proliferation, aged cells may stimulate immune responses that lead to a clearance of aged and damaged cells from tissues.

It is conceivable that oncogenic mutations would stimulate such immune regulatory mechanisms to a greater extent in aged cells than in young ones, resulting in an improved immune clearance of mutant cells during aging. Tumor cells not only need to gain a proliferative advantage but also need to evade immune responses targeting mutant cells that express aberrant protein products (fig.1). The hematopoietic system is not a sessile system but rather a recirculating one, and as leukemia is a disease of bone marrow and blood cells, leukemic cells frequently encounter immune cells on their way through the body. It seems reasonable to assume that molecular mechanisms enabling normal bone marrow/blood/immune cells to circulate are shared also with leukemic cells. Prof. Weissman presented new data from his group showing that the CD47 surface antigen is differentially expressed on resting versus mobilized HSCs in mice and humans. In addition, the surface protein was up-regulated in AML cells compared to non-transformed hematopoietic cells [17, 18].

CD47 is an integrin-associated extracellular ligand for signal-regulatory proteins (SIRPs) [19] that are expressed on macrophages, monocytes, granulocytes and dendritic cells of the myeloid lineage. These cells have phagocytic capabilities and are responsible for the clearance of aged or damaged cells from the blood stream and tissues. The activation of SIRPs leads to the initiation of a signaling cascade resulting in phagocytosis. The interaction of CD47 with SIRPs negatively regulates this signaling and is thought to act as a 'don't eat me' signal. Several studies have shown that the absence of CD47 results in enhanced phagocytosis [20–22].

Prof. Weissman presented experimental data from his group showing an up-regulation of CD47 following mobilization of mouse HSCs into the circulation via cyclophosphamide/G-CSF or lipopolysaccharide treatment. Experiments on CD47 knockout mice demonstrated that CD47 is not necessary for cell migration itself, but rather for the engraftment of mobilized hematopoietic cells in the respective niches, as CD47-null cells were cleared from the circulation by macrophages before they reached their destination.

In a myeloid leukemia mouse model *(hMRP-8bcr/abl × hMRP8bcl2)* [23], CD47 expression was increased 3- to 20-fold on leukemic cells when compared to control mice without leukemia. CD47 expression was also found to be increased in 2 other mouse leukemia models as well as in hematoleukemic cells that were transplanted in wild-type recipient mice. In addition, HSCs and hematopoietic progenitor cells from human AML patients as well as blast crisis cells from CML patients displayed elevated levels of CD47. Moreover, this increase in CD47 expression represented a prognostic marker indicating reduced survival in AML patients [18].

The idea that macrophages phagocytose $CD47^{low/-}$ cells was also tested with cells from the AML-derived human cell line MOLM-13 exhibiting a low endogenous CD47 expression. Untransduced and control vector-transduced MOLM-13 cells were not able to engraft recipient mice. In contrast, retroviral expression of mouse CD47 enabled MOLM-13 cells to engraft recipient mice, resulting in leukemia formation. Interestingly, control vector-transduced or untransduced MOLM-13 cells could also engraft and induce leukemia formation when transplanted in recipient mice that were macrophage-depleted. Co-culture experiments of MOLM-13 cells differing in their CD47 expression levels and macrophages confirmed that CD47 expression rendered the cells resistant to phagocytosis. Moreover, neutralizing antibodies against CD47 or SIRPα (the interaction partner of CD47 on macrophages) enabled phagocytosis of CD47-expressing AML cells by macrophages in co-culture experiments [18]. Experiments of transplanted MOLM-13 cells confirmed that low levels of CD47 expression resulted in the clearance of leukemic cells by macrophages in vivo [17].

In the same studies, transplanted human AML cells with high levels of CD47 expression did not lead to leukemia development in recipient mice when these cells were pre-coated with the blocking anti-CD47 antibody or when the recipient mice were treated with an anti-CD47 antibody after transplantation of the leukemic cells. Secondary transplantation of bone marrow cells from CD47-treated primary recipient mice did not show engraftment of leukemia compared to isotype control-treated mice, suggesting that the AML-initiating or leukemia stem cells can also be targeted with blocking anti-CD47 antibody.

As CD47 is also expressed on normal cells, the question of potential side effects of this treatment remains. In a series of pilot experiments, Prof. Weissman and colleagues could show that anti-CD47 treatment did not deplete endogenous HSCs or confer side effects in wild-type mice [18].

Patterns of Cell Divisions in Leukemia

In addition to evasion of the immune system by expression of the 'don't eat me' signal, CD47 leads to a higher tumorigenic potential of leukemia cells [17], and the dysregulation of cell division patterns and self renewal can lead to more aggressive forms of leukemia. In the laboratory of Dr. Tannishtha Reya, scientists focus on the choice between stem cell renewal and commitment to define how the same signals are subverted in cancer.

Dr. Reya and colleagues have shown through live (time-lapse) imaging that, when transduced into HSCs, the oncogene NUP98-HOXA9, a chromosomal translocation associated with AML and the blast crisis phase of CML, leads to a shift of the normal ratio between asymmetrical and symmetrical divisions in favor of symmetrical divisions [24]. BCR-ABL on the other hand, which is an oncogene associated with chronic-phase CML, does not affect the balance between asymmetrical and symmetrical divisions in HSCs. This may explain why (NUP98-HOXA9-associated) aggressive acute leukemias show a block in differentiation and the leukemic cell mass consists mainly of immature precursor cells, whereas (BCR-ABL-associated) chronic-phase leukemias are slow growing and show normal levels of differentiation.

Together, the above-discussed presentations at the symposium showed that cell-intrinsic and cell-extrinsic alterations can influence the initiation and progression of hematopoietic malignancies. In addition, there is emerging evidence that age-associated alterations in hematopoietic cells can influence these processes. A detailed understanding of molecular alterations in aging hematopoietic cells and the hematopoietic environment should in the future lead to the improvement of therapeutic strategies aiming to treat or prevent the evolution of hematopoietic malignancies.

References

1 Hanahan D, Weinberg RA: The hallmarks of cancer. Cell 2000;100:57–70.
2 Rossi DJ, Jamieson CH, Weissman IL: Stems cells and the pathways to aging and cancer. Cell 2008;132:681–696.
3 Nowell P, Hungerford D: A minute chromosome in human chronic granulocytic leukemia. Science 1960;132:1497.
4 Rowley JD: Letter: a new consistent chromosomal abnormality in chronic myelogenous leukaemia identified by quinacrine fluorescence and Giemsa staining. Nature 1973;243:290–293.
5 Cortez D, Kadlec L, Pendergast AM: Structural and signaling requirements for BCR-ABL-mediated transformation and inhibition of apoptosis. Mol Cell Biol 1995;15:5531–5541.

6 Takahashi N, Miura I, Saitoh K, Miura AB: Lineage involvement of stem cells bearing the Philadelphia chromosome in chronic myeloid leukemia in the chronic phase as shown by a combination of fluorescence-activated cell sorting and fluorescence in situ hybridization. Blood 1998;92:4758–4763.

7 Holyoake T, Jiang X, Eaves C, Eaves A: Isolation of a highly quiescent subpopulation of primitive leukemic cells in chronic myeloid leukemia. Blood 1999; 94:2056–2064.

8 Cobaleda C, Sanchez-Garcia I: B-cell acute lymphoblastic leukaemia: towards understanding its cellular origin. Bioessays 2009;31:600–609.

9 Voncken JW, Kaartinen V, Pattengale PK, Germeraad WT, Groffen J, Heisterkamp N: BCR/ABL P210 and P190 cause distinct leukemia in transgenic mice. Blood 1995;86:4603–4611.

10 Pear WS, Miller JP, Xu L, Pui JC, Soffer B, Quackenbush RC, Pendergast AM, Bronson R, Aster JC, Scott ML, Baltimore D: Efficient and rapid induction of a chronic myelogenous leukemia-like myeloproliferative disease in mice receiving P210 bcr/abl-transduced bone marrow. Blood 1998;92:3780–3792.

11 Bilousova G, Marusyk A, Porter CC, Cardiff RD, DeGregori J: Impaired DNA replication within progenitor cell pools promotes leukemogenesis. PLoS Biol 2005;3:e401.

12 Marusyk A, Casas-Selves M, Henry CJ, Zaberezhnyy V, Klawitter J, Christians U, DeGregori J: Irradiation alters selection for oncogenic mutations in hematopoietic progenitors. Cancer Res 2009; 69:7262–7269.

13 Nalapareddy K, Jiang H, Guachalla Gutierrez LM, Rudolph KL: Determining the influence of telomere dysfunction and DNA damage on stem and progenitor cell aging: what markers can we use? Exp Gerontol 2008;43:998–1004.

14 Marusyk A, DeGregori J: Declining cellular fitness with age promotes cancer initiation by selecting for adaptive oncogenic mutations. Biochim Biophys Acta 2008;1785:1–11.

15 Signer RA, Montecino-Rodriguez E, Witte ON, McLaughlin J, Dorshkind K: Age-related defects in B lymphopoiesis underlie the myeloid dominance of adult leukemia. Blood 2007;110:1831–1839.

16 Signer RA, Montecino-Rodriguez E, Witte ON, Dorshkind K: Aging and cancer resistance in lymphoid progenitors are linked processes conferred by p16Ink4a and Arf. Genes Dev 2008;22: 3115–3120.

17 Jaiswal S, Jamieson CH, Pang WW, Park CY, Chao MP, Majeti R, Traver D, van Rooijen N, Weissman IL: CD47 is upregulated on circulating hematopoietic stem cells and leukemia cells to avoid phagocytosis. Cell 2009;138:271–285.

18 Majeti R, Chao MP, Alizadeh AA, Pang WW, Jaiswal S, Gibbs KD Jr, van Rooijen N, Weissman IL: CD47 is an adverse prognostic factor and therapeutic antibody target on human acute myeloid leukemia stem cells. Cell 2009;138: 286–299.

19 Seiffert M, Cant C, Chen Z, Rappold I, Brugger W, Kanz L, Brown EJ, Ullrich A, Buhring HJ: Human signal-regulatory protein is expressed on normal, but not on subsets of leukemic myeloid cells and mediates cellular adhesion involving its counterreceptor CD47. Blood 1999;94:3633–3643.

20 Brown EJ, Frazier WA: Integrin-associated protein (CD47) and its ligands. Trends Cell Biol. 2001;11:130–135.

21 Blazar BR, Lindberg FP, Ingulli E, Panoskaltsis-Mortari A, Oldenborg PA, Iizuka K, Yokoyama WM, Taylor PA: CD47 (integrin-associated protein) engagement of dendritic cell and macrophage counterreceptors is required to prevent the clearance of donor lymphohematopoietic cells. J Exp Med 2001; 194:541–549.

22 Oldenborg PA, Zheleznyak A, Fang YF, Lagenaur CF, Gresham HD, Lindberg FP: Role of CD47 as a marker of self on red blood cells. Science 2000;288:2051–2054.

23 Jaiswal S, Traver D, Miyamoto T, Akashi K, Lagasse E, Weissman IL: Expression of BCR/ABL and BCL-2 in myeloid progenitors leads to myeloid leukemias. Proc Natl Acad Sci USA 2003; 100:10002–10007.

24 Wu M, Kwon HY, Rattis F, Blum J, Zhao C, Ashkenazi R, Jackson TL, Gaiano N, Oliver T, Reya T: Imaging hematopoietic precursor division in real time. Cell Stem Cell 2007;1:541–554.

Dipl. biol. Anne Gompf
Institute of Molecular Medicine and
Max Planck Research Group on Stem Cell Aging, University of Ulm
Albert-Einstein-Allee 11
DE–89081 Ulm (Germany)
Tel. +49 731 5036 114, Fax +49 731 5036 102
E-Mail anne.gompf@uni-ulm.de

Rudolph KL (ed): Molecular Mechanisms of Adult Stem Cell Aging.
Else Kröner-Fresenius Symp. Basel, Karger, 2010, vol 1, pp 63–78

Intestinal Stem Cell Aging

Tobias Sperka • Karl Lenhard Rudolph

Institute of Molecular Medicine and Max Planck Research Group on Stem Cell Aging, University of Ulm, Ulm, Germany

Abstract

The intestinal epithelium is one of the most rapidly renewing tissues in the human organism. This remarkably high capacity of regeneration requires the maintenance of functional stem cells. Intestinal stem cells are located in the crypts of Lieberkühn, the basal intestinal crypts. In this chapter, we summarize current knowledge on intestinal stem cells. In addition, we summarize reports from the Else Kröner-Fresenius Symposium on the Molecular Mechanisms of Adult Stem Cell Aging indicating that molecular alterations in intestinal stem cells occur during aging and can limit maintenance and function of the intestine.

Intestinal Epithelium

Digestion and resorption of the molecular components of food are the major functions of the intestine. For its task, the small intestine is provided with a special structure consisting of folds, villi, microvilli and invaginations – called the crypts of Lieberkühn – that increase the absorptive surface several times (fig. 1a). Each villus is composed of an epithelial monolayer covering a connective tissue framework that also contains blood and lymphatic vessels. The differentiated cells of the villus epithelium are the enterocytes (which absorb nutrients), goblet cells (which produce a protective mucous) and the enteroendocrine cells (which produce hormones regulating gastrointestinal motility, blood flow or digestion). The differentiated cells in the small intestinal crypt are the Paneth cells, which play a role in host defense by producing bactericidal enzymes such as lysozyme and peptides such as defensins. In addition, the crypts contain intestinal stem cells and their progeny, transient amplifying cells that give rise to the differentiated cells. The main function of the colon is the absorption of water, which is not dependent on such an enlarged surface and therefore contains no villi but only crypts.

Being the direct interface between the nutriment, with all its potentially noxious contaminations, and the body, the cells of the epithelium have to be faithfully maintained to provide a secure barrier and guarantee the body's homeostasis. This is achieved by the constant replacement of exhausted cells with newly formed undamaged progeny originating from the crypt stem cells (fig. 1b). Transient amplifying cells, the direct offspring of the stem cell, produce roughly 300 daughter cells per day per crypt. These progenitor cells are highly proliferative and divide every

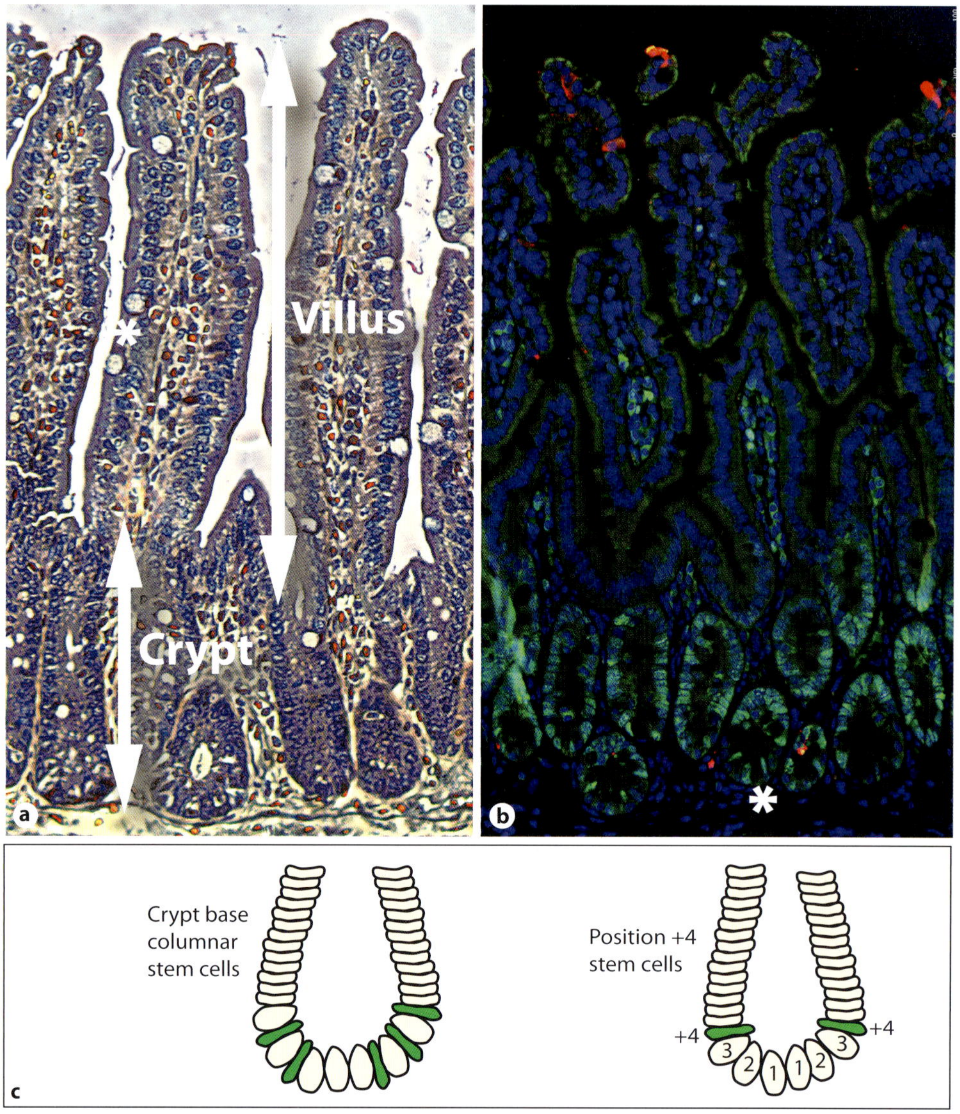

Fig. 1. Architecture of the small intestine. **a** The hematoxylin and eosin stain reveals the principal building units of the mouse small intestine, basal crypts and luminal villi. The villus is formed by an outer epithelial layer made of differentiated cells, for example the prominent mucous secreting goblet cells (asterisk), and contains a stromal core. Each villus is further connected to several crypts demonstrating the continuous nature of the epithelial layer. **b** Basal crypts of Lieberkühn are the major sites of proliferation as revealed by the proliferation marker PCNA (green). They contain dividing stem cells and transient amplifying cells. The crypt base columnar stem cells (asterisk) are readily visible as narrow cone-shaped cells at the base of the crypt interspersed with quiescent Paneth cells. Shedding of exhausted cells at the tip of the villus is executed by apoptosis indicated by cleavage of caspase 3 (red). This spatial separation of proliferation and apoptosis already suggests the cellular flow from the crypt to the villus tip. Nuclei are stained with DAPI (blue). **c** Location of stem cells according to the stem cell zone model (left) and the position +4 model (right).

12–16 h. The progenitor cells reside for only 2–3 days in the crypts before they terminally differentiate [1]. Enterocytes represent the majority of differentiated cells in the murine intestinal epithelium. Enterocytes have a short lifespan and are replenished every 5 days. Together, the intestinal epithelium is one of the fastest renewing tissues in mammals.

Intestinal Stem Cells

Each intestinal crypt is thought to harbor 4–6 intestinal stem cells, but recent data on chromosomal instability and mitochondrial mutations indicate that the true number of stem cells per crypt might be lower. Of note, studies on telomere-dysfunctional mice showed that intestinal loss of p53 cooperates with telomere dysfunction to induce clonal chromosomal aberration in individual crypts. The data indicate that the true number of stem cells per crypt might be lower than originally thought since the results suggest that clonal aberrations originating from 1 or 2 cells occurred in intestinal crypts of these mice. Similarly, data on the propagation of mitochondrial mutations suggest a clonal origin and a smaller number of stem cells per crypt [2, 3]. There are 2 schools of thought on the nature and position of the intestinal stem cell: one favors the +4 position in the crypt (fig. 1c), counting the Paneth cells from the crypt bottom and placing the stem cell between the Paneth and transient amplifying cells [4]. The other – the stem cell zone model – places the stem cells at the crypt bottom between the Paneth cells [5].

Both models have received strong support from lineage-tracing experiments performed in mice. Clevers and colleagues [6] have identified the orphan receptor Lgr5 (leucine-rich-repeat-containing G-protein-coupled receptor 5) as a marker for intestinal stem cells in the stem cell zone but not the +4 position. Lineage-tracing experiments confirmed that the Lgr5-expressing crypt base columnar cells give rise to all 4 differentiated cell types of the small intestine [6]. Sangiorgi and Capecchi [7] employed a very similar strategy investigating Bmi1-expressing cells, which are mainly located in the +4 position. Bmi1 is a subunit of the polycomb-repressing complex 1 and thereby supports heterochromatin formation (see chapters 2 and 8). Deletion of the Bmi1-positive cells led to the loss of crypts of Lieberkühn and a dramatic small intestinal atrophy. Moreover, lineage-tracing experiments (fig. 2) revealed that Bmi1-positive cells give rise to the 4 differentiated cell lineages of the small intestine, while there is no expression of Bmi1 in the colon. In summary, Bmi1 and Lgr5 mark 2 cell fractions at the crypt bottom that harbor stem cell capacity. So far it is not clear if they represent 2 different stages in a differentiation hierarchy, if they trace back to 1 common ancestor or if there are several types of stem cells in the intestine. The first results from mice have demonstrated an overlap of Bmi1-positive and Lgr5-positive cells [8].

Given the highly proliferative renewal of the intestinal epithelium, the maintenance of functional stem cells appears to be of utmost importance to ensure functional integrity of the intestine during aging. Several mechanisms could be involved in intestinal stem cell protection:

(1) Intestinal stem cells reside in an anatomically protected site, the crypt, which is even further occluded by copious amounts of mucous, greatly reducing exposure to potentially toxic luminal content.

(2) The ABC transporter P-glycoprotein/MDR1 is transiently up-regulated in the crypt and villus during regeneration after ischemia-reperfusion damage [9]. Stem cells in various organ systems have been shown to express high levels of multi-drug resistance proteins (i.e. ATP-binding cassette – ABC transporters), likely increasing the resistance of stem cells to the exposure to toxic substances.

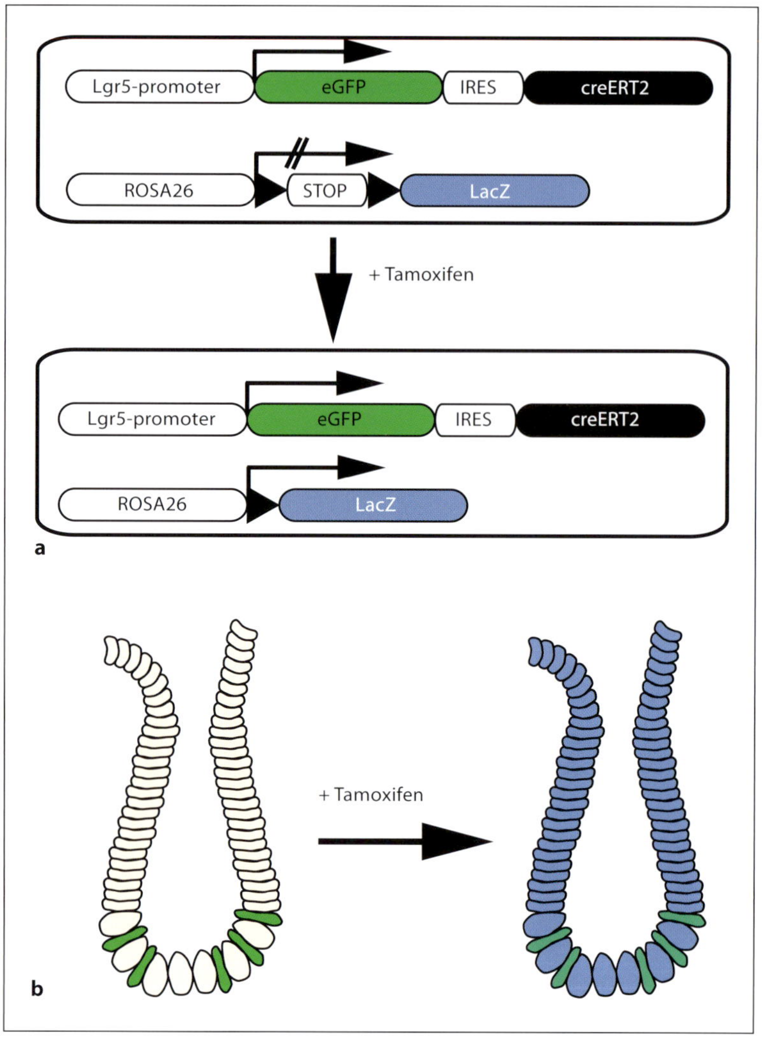

Fig. 2. Lineage tracing in the intestine. **a** A molecular marker (enhanced green fluorescent protein, eGFP) and Cre recombinase (creERT2, fusion of Cre and the estrogen receptor, which is activated upon treatment with tamoxifen) are linked via an internal ribosomal entry site (IRES) and expressed under the control of a stem cell-specific promoter (e.g. Lgr5). Expression of a second molecular mark (e.g. LacZ) is blocked by a STOP codon flanked by loxP sites (black triangles). Addition of tamoxifen renders Cre recombinase active and excises the STOP codon allowing LacZ expression. **b** Addition of tamoxifen induces LacZ expression in all Lgr5 cells and their cellular progeny (turquoise and blue stain, respectively). With Lgr5 cells being stem cells, their progeny will permanently mark the whole crypt and villus – non-stem cells would only give rise to transient and patchy marking.

(3) There is evidence that an asymmetric segregation of damaged proteins occurs during stem cell division, keeping stem cells free of damaged proteins [10]. Such mechanisms of asymmetric segregation of proteins have also been described in yeast cell division [11]. It is tempting to speculate that this mechanism is also active in intestinal stem cells.

(4) Intestinal stem cells are much more sensitive to ionizing γ-irradiation than the transient amplifying cells [6]. The induction of apoptosis rather than cell cycle arrest and damage repair could be interpreted as the avoidance of the error-prone DNA repair mechanisms that can lead to fatal mutations and cancer formation [4]. On the other hand, it is questionable whether increased damage sensitivity would really be advantageous at the tissue level since stem cell depletion would ultimately lead to a loss of the critical cell compartment that is required for tissue repair. In fact, there is evidence that hematopoietic cells may exhibit a higher level of damage tol-

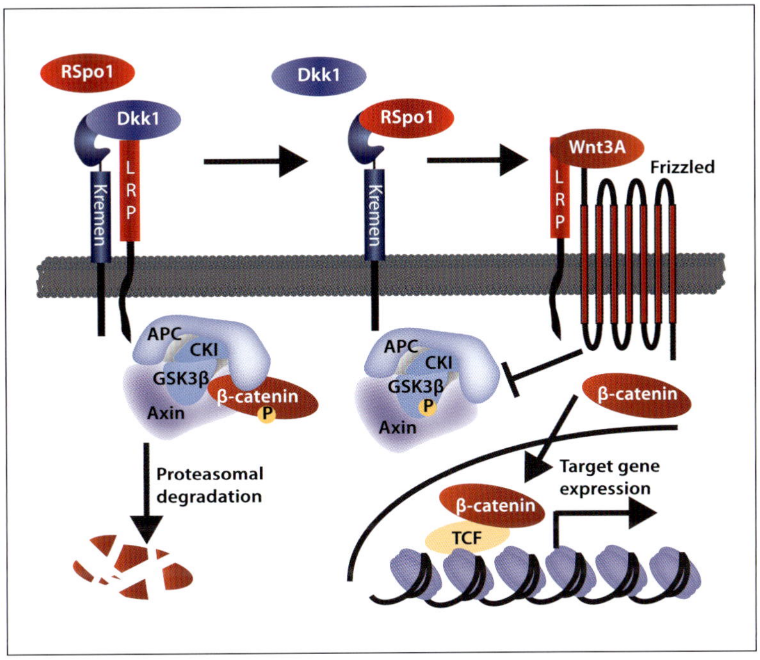

Fig. 3. Wnt pathway. Binding of Wnt3A to the co-receptor LRP and the Frizzled receptor inhibits the intracellular destruction complex composed of APC, Axin, CKI and phosphorylated GSK3β inducing cytosolic β-catenin levels to rise (right). β-catenin translocates to the nucleus and activates TCF transcription factors enhancing target gene expression (right). β-catenin is earmarked for proteasomal destruction in the absence of Wnt activity through phosphorylation (P) via the destruction complex (left). Wnt signaling is further inhibited through formation of a complex between the LRP co-receptor and the inhibitors Dkk1 and Kremen, which can be released via R-spondin1 (RSpo1) (left; refer to text for further details).

eration compared to progenitor cells to avoid depletion of stem cells [12].

(5) The rapidly renewing intestine was the first organ for which asymmetric DNA strand allocation was described. The rationale behind this concept, initially proposed in 1975 [13], indicates that the maintenance of an immortal, unreplicated DNA strand in stem cells protects stem cells against replication-induced DNA mutations.

Molecular Control of Cell Proliferation and Organ Maintenance in the Intestine

The primary mechanism inducing intestinal crypt cell proliferation is the canonical Wnt signaling pathway (fig. 3). Under resting conditions, β-catenin is phosphorylated and targeted for proteasomal degradation. Phosphorylation is accomplished by the 2 kinases GSK3β (glycogen synthase kinase 3β) and CKI (casein kinase I) as part of a destruction complex also containing the tumor suppressors axin and Apc (adenomatous polyposis coli). Upon binding of Wnt ligand to the Frizzled receptor, the destruction complex is inhibited leading to β-catenin stabilization and its transport to the nucleus, where it serves as a co-activator of TCF/LEF transcription factors (T cell factor/lymphocyte enhancer factor). Several genetic studies have demonstrated the pivotal role of the canonical Wnt signaling pathways in intestinal maintenance [14, 15]. In addition, there is evidence that the JNK pathway can accelerate the activity of the Wnt signaling pathway, resulting in increased proliferation in the stem cell and transient amplifying cell compartment [16]. These data indicated that both canonical and non-canonical Wnt signaling influence the size of the intestinal progenitor cell compartment.

The Wnt signaling pathway has also been implicated in the control of stem cell differentiation in the intestine (fig. 4). The first level of lineage separation is the division into absorptive (enterocytes) and secretory cells (goblet, Paneth and en-

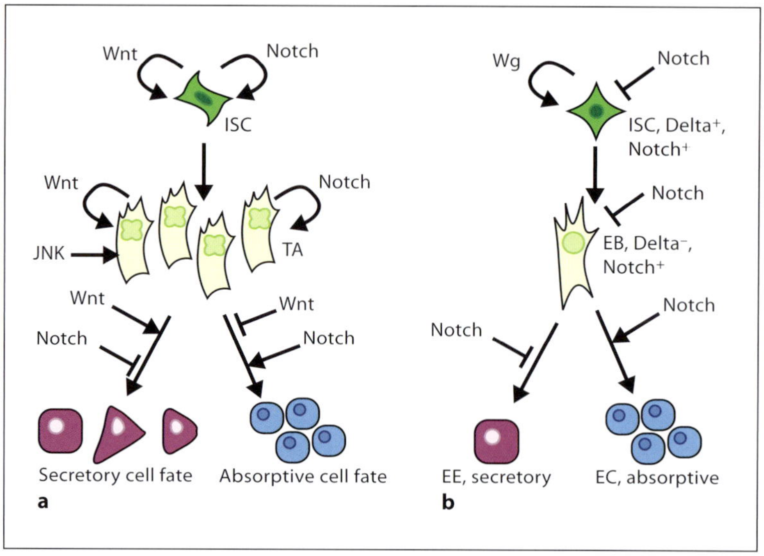

Fig. 4. Genetic hierarchy of epithelial lineage commitment in the intestine. The differentiation hierarchies of mammalia (illustrated by mouse in **a**) and *D. melanogaster* (**b**) show striking similarities and some differences. **a** Intestinal stem cells (ISC) give rise to a transient amplifying (TA) pool of cycling progenitor cells, which differentiate into absorptive or secretory cells. Wnt and Notch signaling have been shown to stimulate proliferation as well as differentiation in the mouse. JNK has been shown to enlarge the cycling TA compartment. **b** The *D. melanogaster* intestine displays a similar hierarchy with ISCs giving rise to quiescent enteroblasts (EB), which differentiate into enteroendocrine cells (EE) and enterocytes (EC). Wingless (Wg) stimulates ISC proliferation and Notch blocks the expansion of the proliferating ISC pool. Notch specifies an absorptive cell fate.

teroendocrine cells). Enhanced Wnt signaling blocks normal differentiation into enterocytes and supports formation of Paneth cells [17]. In contrast, loss of the Wnt target Tcf-4 enhances differentiation into enterocytes but blocks enteroendocrine differentiation [14]. Together, these results suggest that the Wnt pathway specifies a secretory lineage decision in intestinal stem cells [18].

The Notch pathway has been shown to be another fate-specifying route, since Notch led to an enhanced generation of secretory cells and a reduction of proliferative crypt cells [19–22]. Induction of intestinal cell proliferation in response to Notch activation is thought to involve the derepression of the cell cycle inhibitors p27 and p57 [23]. Interestingly, the Wnt and Notch pathways are both involved in self-renewal of the stem cell compartment and specify differentiation into either the secretory or the absorptive lineage, respectively (fig. 4). This creates an obvious need for a lineage commitment along the crypt, starting already at the stem cell level. Whereas Wnt growth factors are secreted diffusible ligands, Notch signaling is activated through the ligands Delta and Jagged, which are presented from a neighboring cell. Notch signaling is further characterized by an incompletely understood process called lateral inhibition: a population of Notch and Delta double-positive cells develops such that only single-positive cells remain. This has lead to the suggestion that intestinal Wnt$^+$Notch$^+$ cells divide until, through lateral inhibition, Wnt$^+$Notch$^-$ cells develop, which turn into secretory cells. However, the size of the group of Wnt-activated cells is limited as a result of the short- and long-range spatial signals (Wnt ligand diffusion). This would create $Wnt^{-}Notch^{+}$ cells committed to become absorptive tissue, and $Wnt^{-}Notch^{-}$ cells taking the secretory route [24]. The balance between Notch and JNK signaling has recently been found to be altered in aging flies, leading to severe changes in intestinal tissue architecture [25].

Age-Associated Diseases of the Intestine

A variety of age-associated disorders and diseases have been linked to intestinal functions, although it is still unclear whether the evolution of disease pathology involves an age-dependent dysfunction of intestinal stem cells.

Intestinal Motor Dysfunction
Age-related intestinal motor dysfunction particularly affects the large intestine, where it manifests itself as an increase in transit time in old age. As a result, water resorption increases, leading to obstipation. Age-related changes in neurons and receptors are responsible for the disease [26]. The role of enteroendocrine cells secreting several peptide hormones, which are involved in motor control, remains to be defined.

Weight Loss
A Canadian survey determined that 59% of elderly patients were malnourished or at risk or malnourishment, which may be caused by an age-related decline in absorptive function of the intestine, among other factors [27]. The proliferating zone in the crypt is markedly broadened in old humans and proliferation marker-positive enterocytes are detected in the villus of old individuals, suggesting the presence of altered differentiation patterns and potentially impaired absorptive function in the elderly [28]. Additionally, it has been demonstrated that the aged intestine shows a reduced adaptive response under conditions of stress. Severe dietary restriction in aged rats induced a dramatic weight loss compared to young rats [29]. The intestine of these animals showed progressive atrophy and ileal hydrolase activity was decreased. Following a period of stress, such as illness or injury, it has been shown in humans that elderly patients continued to underfeed themselves while younger patients increased their energy intake [30]. In elderly patients, the intestine may have a decreased functional reserve and so they may become undernourished more rapidly during acute phases of stress, thus requiring an extended period of intensive nutritional care [26]. The role of intestinal failure in normal aging remains under debate. While there is no clear evidence of an age-related atrophy of the intestinal epithelium [31], functional impairments in absorption of nutrients appear to represent a clinically relevant problem in the elderly [32].

Diabetes
Age-associated type 2 diabetes is a widespread health problem characterized by elevated blood glucose levels, impaired insulin responsiveness, dysfunctional pancreatic β cells and a reduced incretin hormone effect. This effect provokes a greater insulin release from β cells upon intestinally administered glucose than the same amount of glucose given intravenously. This is explained by the response of enteroendocrine cells producing various hormones upon glucose stimulation (e.g. GLP-1), which act directly on pancreatic β cells to stimulate insulin secretion. The incretin hormones thus have blood glucose-lowering effects and so have gained clinical interest [33]. GLP-1 expression is regulated via the Wnt target gene Tcf4/TCF7L2, and humans with a genetic variant of TCF7L2 displaying an impaired incretin effect though reduced levels of serum GLP-1 were not detected [34]. Furthermore, the balance between Notch and Wnt signaling, which controls stem cell differentiation into secretory and absorptive epithelia [18], seems to be disturbed upon aging, as recently shown in the

Drosophila gut [25]. It is tempting to speculate that aging-associated changes in signaling pathways that control differentiation, and hence the expression of the enteroendocrine secretome, impact on a wide variety of metabolic processes, including diabetes.

Ischemia Reperfusion Damage

A partial or complete blocking of the intestinal blood supply triggered by low blood pressure or a thrombus and the subsequent ischemia reperfusion damage cause ischemic colitis. The incidence of ischemic colitis is strongly age-correlated, with 90% of affected individuals being over 60 years of age [35] and complications affecting mainly the elderly [36]. Complete intestinal repair is achieved by enhanced crypt proliferation and differentiation [37–39]. Analyses in humans, rodents and flies have shown that proliferation and differentiation become unbalanced in the aged intestine [25, 28, 40–42], but the clinical relevance of this has yet to be determined.

Cancer

Intestinal cancer is the second biggest cancer threat for humans after lung cancer. Colon and rectum are much more prone to cancer development while the small intestine seems to be better protected from neoplastic transformation. Cancer and aging are also well correlated, which can be subdivided into 3 phases: until the end of the fourth decade of life there are relatively few cases, from the age of 40 to the end of the eighth decade of life tumor burden rises steadily, but then it drops again significantly from the age of 80 onwards[1].

The accumulation of DNA damage could be important for both cancer and aging. For example, mutations in the DNA mismatch repair (MMR) factors MLH1, MSH2, MSH6 and PMS2 are associated with a hereditary form of colorectal cancer, also called Lynch syndrome [43]. Such MMR deficiency induces high mutation rates, manifested in microsatellite instability, a general form of genomic instability [44]. Additionally, several signaling pathways are affected by mutations that lead to their stimulation or inactivation, including the Wnt, Akt and MAPK pathways [45].

The identification of Lgr5- and Bmi1-labeled cells as the intestinal stem cells together with the generation of mouse models expressing Cre recombinase stem cells specifically allowed the testing of whether intestinal stem cells are the cell type of origin for intestinal cancer development. Both studies showed that the activation of Wnt/β-catenin signaling in intestinal stem cells led to intestinal adenoma, whereas activation of this pathway in differentiated enterocytes did not result in adenoma formation [7, 46]. Together, these data support the concept that tumors arise from tissue stem cells.

Mechanisms of Intestinal Stem Cell Aging in Model Systems

The analysis of model systems has increased our understanding of the aging process. Two different models to study intestinal stem cell aging were presented at the Else Kröner-Fresenius Symposium on the Molecular Mechanisms of Adult Stem Cell Aging: the telomerase knockout mice (mTerc$^{-/-}$) and the fly *Drosophila melanogaster*.

Critically short telomeres cause activation of the DNA damage response and subsequently induce cellular senescence or apoptosis, which in stem cells would deplete the active stem cell pool (see chapter 9). Telomere shortening is observed in many organs in humans [47] and has been correlated with increased mortality rates from infectious and heart diseases [48]. Moreover, there is genetic evidence for a role of telomere shortening in human aging. Patients with some subtypes of

[1] Source: Center of Disease Control WONDER database, which covers United States cancer statistics from 1999 until 2004.

Dyskeratosis congenita show rapid telomere loss due to a mutation in the RNA component hTERC. These patients suffer from bone marrow and intestinal failure as well as from an accelerated cancer incidence. Together, these data indicate that human telomeres are not unlimited and telomere shortening can contribute to organ failure.

Studies on mTerc$^{-/-}$ mice have provided experimental proof that telomere shortening can lead to premature atrophy in high-turnover organs [49]. Deletion of mTerc renders the telomerase reverse transcriptase inactive with respect to telomere extension. In this model, mTerc loss induces many features of Dyskeratosis congenita and human aging, including impaired lymphopoiesis (see chapter 3), intestinal atrophy, hair graying and loss, delayed wound healing and an elevated cancer incidence [50]. Remarkably, all described phenotypes occur only in late-generation knockout mice, indicating that the long mouse telomeres provide enough reserve to protect the organism in early generations without telomerase activity. Subsequent cell divisions erode the telomeres to a critically short length in later generations, which induces senescence and apoptosis leading to organ failure and aging. This late development of the phenotype argues for a direct causal link between telomerase dysfunction, telomere shortening and aging: the loss of telomerase activity does not directly lead to the observed phenotypes but only after telomere attrition, supporting the idea that telomere shortening in humans is a direct cause of aging.

Mitotically active cells encounter constant telomere shortening, which has to be balanced by telomere synthesis. The intestinal crypts display long telomeres [51] enabled by telomerase expression [52, 53]. Late-generation mTerc knockout mice with dysfunctional telomeres display a severe gut atrophy combined with weight loss [50]. Telomere dysfunction has also been associated with increased radiosensitivity of mouse gut, demonstrating the essential role of telomeres in gut homeostasis and regeneration in response to irradiation-induced injury [54], which may have clinical implications for radiotherapy in old patients [55].

Furthermore, telomere dysfunction itself induces a DNA damage response mediated by classical DNA damage checkpoint proteins including p53 and pRb, which causes cell cycle arrest, DNA repair, senescence and apoptosis [56]. In vitro studies on human cells have shown that abrogation of these checkpoints led to an elongated proliferative lifespan of human fibroblasts [57], which prompted mouse in vivo studies to test the relevance of these pathways in stem cell aging.

Combined deletion of mTerc and p21 – a pivotal downstream effector of the p53-induced senescence response [58] – improved overall fitness and extended the lifespan of telomere-dysfunctional mice [59]. This rescue was associated with an improved maintenance of intestinal crypts and villus formation preventing the premature weight loss in aging mTerc$^{-/-}$ mice compared to wild-type mice. p21 deletion did not lead to telomere repair, but stem cells with moderate levels of telomere dysfunction retained the capacity to regenerate the intestine to a functional level. In addition, these mice did not show increased cancer development, indicating that the tumor-suppressive and aging-promoting activities of p53 are separable at the level of downstream targets. These results indicate that a functional p53-induced apoptosis response may be sufficient to protect from carcinogenesis in mTerc$^{-/-}$ mice and warrant further studies by combining the mTerc knockout with known p53 apoptosis mediators like Puma and Noxa [60].

At the symposium, new data were presented pointing to a tissue-protective role of p53 in telomere-dysfunctional tissues. In contrast to the data on p21, deletion of p53 did not rescue intestinal atrophy in aging telomere-dysfunctional mice [2]. Accelerated tissue atrophy in compound mutant mice was associated with the clonal evolution of chromosomal instability, indicating that p53 deletion led to the aberrant survival of

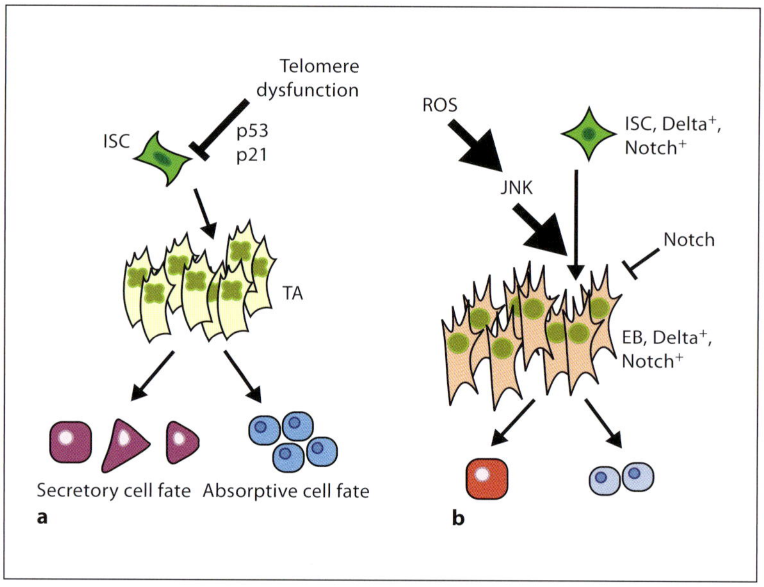

Fig. 5. Age-associated changes in mammalia (**a**) and *D. melanogaster* (**b**). **a** The TA compartment is enlarged in aging humans, the effects on differentiation are only moderate but still not fully clarified. Telomere dysfunction induces a loss of the stem cell compartment, which involves activation of the p53-p21 pathways. **b** The aging *D. melanogaster* intestine suffers from ROS-induced JNK activation, which produces an enlarged and incorrectly differentiated EB population. Note the prevailing Delta expression in the EBs, which is not detectable in the young intestine, and which causes incorrectly differentiated EB progeny. Genetic reduction of Delta expression reverts the incorrect differentiation.

chromosomally unstable intestinal stem cells in aging telomere-dysfunctional mice. Dr. Karl Lenhard Rudolph concluded from these data that p53 protects tissue from atrophy in the context of telomere dysfunction and aging by depleting chromosomally unstable stem cells. The aberrant survival of chromosomally unstable intestinal stem cells resulted in aberrant differentiation, activation of p53-independent apoptosis at progenitor cell level, and accelerated development of tissue atrophy (fig. 5).

Several other studies investigated the effects of compromising DNA repair pathways in the *mTerc*$^{-/-}$ mouse, thus testing for their influence in the response to dysfunctional telomeres. Atm kinase is activated in response to deprotected telomeres and activates p53 [61]. Surprisingly, Atm deletion did not rescue premature aging and intestinal atrophy in telomere-dysfunctional mice and p53 was still activated, potentially through mechanisms involving Atr, another DNA damage response kinase [61, 62]. Further upstream in the response cascade after telomere deprotection are various DNA binding and processing factors like Exo1 (exonuclease-1), which is known to be involved in the senescence response towards short telomeres in yeast [63]. Compound deletion of Terc and Exo1 improved the fitness of telomere-dysfunctional mice, including elongated lifespan and delayed intestinal atrophy characterized by reduced apoptosis and genomic instability in intestinal crypts [64]. As with the mTerc$^{-/-}$, p21$^{-/-}$ mice, Exo1 deletion in mTerc$^{-/-}$ mice did not rescue telomere dysfunction but prevented accumulation of DNA damage signaling components such as γH2AX, 53BP1, p53 and p21 in intestinal crypts. Furthermore, no increase in chromosomal instability and tumor formation was detected, which could be interpreted as Exo1 being necessary for telomere processing with subsequent chromosomal fusion. This hypothesis has received support from recent findings demonstrating that Exo1 is important for DNA double-strand break processing, which has similarities with the processing of dysfunctional telomeres [65].

Non-homologous end joining (NHEJ) is another route to repair DNA double-strand breaks and supports the fusion of deprotected telomeres

[66]. Telomeric fusions and the subsequent formation of mitotic bridges potentially support the development of chromosomal instability after repetitive breakage-fusion-bridge cycles [67]. The supplementary knockout of the NHEJ factors DNA-PK or Ku86 in telomere-dysfunctional mice led to shortened life-span and worsening of the intestinal atrophy, which was distinguished by reduced apoptosis and proliferation [68, 69]. However, another study that depleted DNA-PK in Terc$^{-/-}$ mice did not reveal any significant influence of the NHEJ pathway on dysfunctional telomere processing and intestinal atrophy [70]. The results presented underline the complex cellular response towards dysfunctional telomeres, and it will be interesting to test other DNA damage response pathways for their contribution to pro-aging and anti-tumor function, which will help elucidate the connections between aging, DNA damage responses and genomic stability.

Free radicals are another form of cellular stress. They cause damage to DNA, proteins and lipids, which is implicated in cellular and organismal aging (see chapter 8). Furthermore, there is a link between free radicals and telomere attrition [71] since the telomeric guanine triplet is very sensitive to oxidative damage-induced cleavage, explaining in part the rapid replicative senescence of certain primary cells under normal oxygen conditions [72]. Sources for reactive oxygen species (ROS) are manifold and include irradiation, toxins, pathogens, oncogenes [73] and normal metabolism. One major source of ROS is the respiratory chain located in the mitochondria. ROS are produced as by-products during ATP generation, although they are normally detoxified [74]. There are several studies that show a functional decline of mitochondria during aging, which manifests in an accelerated increase in mutations of mitochondrial DNA [75, 76] and subsequently elevated ROS production. Studies on aging intestinal crypts have revealed an increase in mutations affecting the respiratory chain component cytochrome c oxidase, though ROS levels have yet to be analyzed [3]. Naturally, there is a plethora of protective antioxidant defense mechanisms [77]. Amongst the pathways responding to oxidative stress [78, 79], the c-Jun amino-terminal kinase JNK has recently been associated with intestinal stem cell aging in *D. melanogaster.*

The discovery of intestinal stem cells in the fly was a surprise since *Drosophila* was considered to be a 'post-mitotic' organism. Intestinal stem cells of the midgut give rise to quiescent daughter cells, called the enteroblasts, which differentiate into enteroendocrine cells and enterocytes [80, 81] (fig. 4). Both of these differentiated populations have a lifetime of about a week before they are replaced. The morphological similarities to the mammalian system prompted analysis of the Notch and Wnt pathways. Intestinal stem cells receive Wingless (Wg, the *Drosophila* homologue of Wnt) signals secreted from neighboring muscle cells [82]. Interference with Wg signaling leads to a small reduction in proliferation and differentiation in intestinal stem cells, indicating a similar – albeit not as prominent – influence of Wg as in the mammalian system. The Notch pathway induces equal but also different phenotypes in the insect and mammalian intestine. Whereas the inhibition of Notch in the murine gut induces progenitor cell depletion and differentiation [21], in *Drosophila* it induces intestinal stem cell over-proliferation due to impaired differentiation [80, 81]. Concordantly, elevated Notch signaling expands the mammalian progenitor pool [83] but induces differentiation in *Drosophila* [80, 81]. The selection between the secretory and absorptive lineage relies on Notch in both flies and mice. Notch inhibition induces secretory fate decision in the murine intestine by turning cells into goblet cells and into enteroendocrine cells in *Drosophila* [81]. According to the observed homologies, the fly intestine may serve as a model for stem cell aging.

During the symposium, Dr. Benoit Biteau from the group of Dr. Heinrich Jasper (Rochester, N.Y., USA) described age-associated changes of the *Drosophila* midgut. He highlighted the overproliferation of intestinal stem cells and the accumulation of their abnormally differentiated daughter cells, which cause deterioration of the intestinal epithelium [25] (fig. 5). The differentiation defect is associated with Notch signaling. In young flies, Delta ligand-expressing intestinal stem cells induce Notch activity in progenitor enteroblasts, which do not express Delta [84]. This characteristic of lineage specification is disturbed upon aging in the form of the accumulation of Notch$^+$/Delta$^+$ daughter cells that are not able to properly differentiate, hence destroying the intestinal architecture. Dr. Biteau reported elevated JNK activity in the aging *Drosophila* intestinal stem cells and enteroblasts as the cause for increased proliferation and elevated Delta expression, leading to the subsequent differentiation failure [25]. It was possible to recapitulate the aging phenotype by addition of the ROS-inducing compound Paraquat, which activates JNK, and to revert the phenotype by inhibiting JNK, demonstrating the causal role played by JNK in age-related tissue degeneration. Recent studies suggest that the production of ROS in response to an increased bacterial population is the reason for the elevated JNK activity in the aged fly gut [85]. The consequences are elevated proliferation and imbalanced differentiation due to improper Notch/Delta signaling, underlining the detrimental effect of deficient lineage specification on tissue homeostasis. A twist to the JNK connection comes from the Jasper laboratory itself, providing evidence that JNK activation leads to an extended lifespan by conferring tolerance to oxidative stress [86]. Too much stress, and hence too much JNK activity, probably tips the balance from life extension towards proliferation and failed differentiation, leading to the loss of tissue integrity observed upon aging.

In vitro Culture System

Another important advancement for the study of intestinal stem cells was the development of an in vitro culture system, which was possible after the identification of the Lgr5 and Bmi1 stem cells. Sustained intestinal renewal requires active Wnt signaling and had to be adapted to the in vitro culture [87]. In vivo studies had shown that Wnt induction via the enhancer R-spondin1 led to an enlarged crypt area and massive proliferation [88, 89]. The use of R-spondin1 in a collagen matrix suffices to support the growth of isolated intestinal organoids for more than 1 year in vitro, leading to the formation of hollow cyst-like structures with crypts on the outside and villi projecting into the lumen [90]. Recapitulating the action in vivo, proliferation is detected in the crypt structures and apoptotic cells are found in the lumen of the cyst. This type of system uses a mixed population of primary material that contains at least stem cells, differentiated epithelia and myofibroblasts, indicating the transfer of a whole stem cell niche into the culture dish. The second in vitro culture system starts from a single Lgr5-positive cell also yielding hollow crypt organoids with the proliferating crypt pointing outward and a differentiated epithelium plus apoptotic cell shedding towards the lumen [91]. Single cells derived from such a single cell-derived crypt organoid are again able to form several organoids emanating from individual cells, indicating the presence of more than 1 stem cell. This impressive achievement of intestinal stem cell self-renewal without a stem cell niche is made possible with the combined use of R-spondin1 in a Matrigel matrix, the activation of EGF and Notch signaling [21, 92] and the inhibition of BMP and ROCK signaling [93–95].

Outlook

The in vitro culture systems will simplify the study of stem cell-driven crypt-villus biology, including renewal, differentiation and aging. Furthermore, they may be used to study interactions with other cell types such as smooth muscle, neurons, endothelium or immune cells, which will help to elucidate the stem cell niche. In addition, they may facilitate studies on host-pathogen interactions, regenerative medicine, gene therapy and cancer. The aging field has benefited from sophisticated aging models that recapitulate the human process and explain it as a combination of loss of stem cell self-renewal, deregulated differentiation and inability to remove dysfunctional stem cells. The study of animal aging models will additionally help elucidate the capabilities of aged stem cells and will show up connections to disease states. Careful analysis has succeeded in dissecting the pathways influencing aging and tumor suppression, showing research avenues that may allow future therapeutic interventions to support healthier aging while keeping up the necessary tumor barriers. A therapy targeting the origin of differentiation, regeneration and proliferation – i.e. the stem cell – would be more likely to treat the cause and not only the symptom of a disorder.

References

1 Marshman E, Booth C, Potten CS: The intestinal epithelial stem cell. Bioessays 2002;24:91–98.
2 Begus-Nahrmann Y, Lechel A, Obenauf AC, Nalapareddy K, Peit E, Hoffmann E, Schlaudraff F, Liss B, Schirmacher P, Kestler H, Danenberg E, Barker N, Clevers H, Speicher MR, Rudolph KL: p53 deletion impairs clearance of chromosomal instable stem cells compromising integrity of the intestine in aging telomere dysfunctional mice. Nat Genet 2009;41:1138–1143.
3 Taylor RW, Barron MJ, Borthwick GM, Gospel A, Chinnery PF, Samuels DC, Taylor GA, Plusa SM, Needham SJ, Greaves LC, Kirkwood TB, Turnbull DM: Mitochondrial DNA mutations in human colonic crypt stem cells. J Clin Invest 2003;112:1351–1360.
4 Potten CS: Extreme sensitivity of some intestinal crypt cells to X and gamma irradiation. Nature 1977;269:518–521.
5 Cheng H, Leblond CP: Origin, differentiation and renewal of the four main epithelial cell types in the mouse small intestine. V. Unitarian theory of the origin of the four epithelial cell types. Am J Anat 1974;141:537–561.
6 Barker N, van Es JH, Kuipers J, Kujala P, van den Born M, Cozijnsen M, Haegebarth A, Korving J, Begthel H, Peters PJ, Clevers H: Identification of stem cells in small intestine and colon by marker gene Lgr5. Nature 2007;449: 1003–1007.
7 Sangiorgi E, Capecchi MR: Bmi1 is expressed in vivo in intestinal stem cells. Nat Genet 2008;40:915–920.
8 van der Flier LG, van Gijn ME, Hatzis P, Kujala P, Haegebarth A, Stange DE, Begthel H, van den Born M, Guryev V, Oving I, van Es JH, Barker N, Peters PJ, van de Wetering M, Clevers H: Transcription factor achaete scute-like 2 controls intestinal stem cell fate. Cell 2009;136:903–912.
9 Omae T, Goto M, Shimomura M, Masuda S, Ito K, Okuda M, Inui K: Transient up-regulation of P-glycoprotein reduces tacrolimus absorption after ischemia-reperfusion injury in rat ileum. Biochem Pharmacol 2005;69:561–568.
10 Rujano MA, Bosveld F, Salomons FA, Dijk F, van Waarde MA, van der Want JJ, de Vos RA, Brunt ER, Sibon OC, Kampinga HH: Polarised asymmetric inheritance of accumulated protein damage in higher eukaryotes. PLoS biology 2006;4:e417.
11 Shcheprova Z, Baldi S, Frei SB, Gonnet G, Barral Y: A mechanism for asymmetric segregation of age during yeast budding. Nature 2008;454:728–734.
12 Rossi DJ, Bryder D, Seita J, Nussenzweig A, Hoeijmakers J, Weissman IL: Deficiencies in DNA damage repair limit the function of haematopoietic stem cells with age. Nature 2007;447: 725–729.
13 Cairns J: Mutation selection and the natural history of cancer. Nature 1975; 255:197–200.
14 Korinek V, Barker N, Moerer P, van Donselaar E, Huls G, Peters PJ, Clevers H: Depletion of epithelial stem-cell compartments in the small intestine of mice lacking Tcf-4. Nat Genet 1998;19:379–383.
15 Pinto D, Gregorieff A, Begthel H, Clevers H: Canonical Wnt signals are essential for homeostasis of the intestinal epithelium. Genes Dev 2003;17:1709–1713.
16 Sancho R, Nateri AS, de Vinuesa AG, Aguilera C, Nye E, Spencer-Dene B, Behrens A: JNK signalling modulates intestinal homeostasis and tumourigenesis in mice. EMBO J 2009;28:1843–1854.
17 Sansom OJ, Reed KR, Hayes AJ, Ireland H, Brinkmann H, Newton IP, Batlle E, Simon-Assmann P, Clevers H, Nathke IS, Clarke AR, Winton DJ: Loss of Apc in vivo immediately perturbs Wnt signaling, differentiation, and migration. Genes Dev 2004;18:1385–1390.
18 van der Flier LG, Clevers H: Stem cells, self-renewal, and differentiation in the intestinal epithelium. Annual Rev Physiol 2009;71:241–260.
19 Jensen J, Pedersen EE, Galante P, Hald J, Heller RS, Ishibashi M, Kageyama R, Guillemot F, Serup P, Madsen OD: Control of endodermal endocrine development by Hes-1. Nat Genet 2000;24:36–44.

20 Milano J, McKay J, Dagenais C, Foster-Brown L, Pognan F, Gadient R, Jacobs RT, Zacco A, Greenberg B, Ciaccio PJ: Modulation of notch processing by gamma-secretase inhibitors causes intestinal goblet cell metaplasia and induction of genes known to specify gut secretory lineage differentiation. Toxicol Sci 2004;82:341–358.
21 van Es JH, van Gijn ME, Riccio O, van den Born M, Vooijs M, Begthel H, Cozijnsen M, Robine S, Winton DJ, Radtke F, Clevers H: Notch/gamma-secretase inhibition turns proliferative cells in intestinal crypts and adenomas into goblet cells. Nature 2005;435:959–963.
22 Wong GT, Manfra D, Poulet FM, Zhang Q, Josien H, Bara T, Engstrom L, Pinzon-Ortiz M, Fine JS, Lee HJ, Zhang L, Higgins GA, Parker EM: Chronic treatment with the gamma-secretase inhibitor LY-411,575 inhibits beta-amyloid peptide production and alters lymphopoiesis and intestinal cell differentiation. J Biol Chem 2004;279:12876–12882.
23 Riccio O, van Gijn ME, Bezdek AC, Pellegrinet L, van Es JH, Zimber-Strobl U, Strobl LJ, Honjo T, Clevers H, Radtke F: Loss of intestinal crypt progenitor cells owing to inactivation of both Notch1 and Notch2 is accompanied by derepression of CDK inhibitors p27Kip1 and p57Kip2. EMBO Rep 2008;9:377–383.
24 Crosnier C, Stamataki D, Lewis J: Organizing cell renewal in the intestine: stem cells, signals and combinatorial control. Nat Rev Genet 2006;7:349–359.
25 Biteau B, Hochmuth CE, Jasper H: JNK activity in somatic stem cells causes loss of tissue homeostasis in the aging *Drosophila* gut. Cell Stem Cell 2008;3:442–455.
26 Salles N: Basic mechanisms of the aging gastrointestinal tract. Dig Dis 2007;25:112–117.
27 Azad N, Murphy J, Amos SS, Toppan J: Nutrition survey in an elderly population following admission to a tertiary care hospital. CMAJ 1999;161:511–515.
28 Corazza GR, Ginaldi L, Quaglione G, Ponzielli F, Vecchio L, Biagi F, Quaglino D: Proliferating cell nuclear antigen expression is increased in small bowel epithelium in the elderly. Mech Ageing Dev 1998;104:1–9.
29 Chambon-Savanovitch C, Felgines C, Farges MC, Pernet P, Cezard J, Raul F, Cynober L, Vasson MP: Severe dietary restriction initiated in aged rats: evidence for poor adaptation in terms of protein metabolism and intestinal functions. Eur J Clin Invest 1999;29:504–511.
30 Roberts SB, Fuss P, Heyman MB, Evans WJ, Tsay R, Rasmussen H, Fiatarone M, Cortiella J, Dallal GE, Young VR: Control of food intake in older men. JAMA 1994;272:1601–1606.
31 Holt PR: Intestinal malabsorption in the elderly. Dig Dis 2007;25:144–150.
32 Woudstra T, Thomson AB: Nutrient absorption and intestinal adaptation with ageing. Best Pract Res 2002;16:1–15.
33 Wideman RD, Kieffer TJ: Mining incretin hormone pathways for novel therapies. Trends Endocrinol Metabol 2009;20:280–286.
34 Lyssenko V, Lupi R, Marchetti P, Del Guerra S, Orho-Melander M, Almgren P, Sjogren M, Ling C, Eriksson KF, Lethagen AL, Mancarella R, Berglund G, Tuomi T, Nilsson P, Del Prato S, Groop L: Mechanisms by which common variants in the TCF7L2 gene increase risk of type 2 diabetes. J Clin Invest 2007;117:2155–2163.
35 Higgins PD, Davis KJ, Laine L: Systematic review: the epidemiology of ischaemic colitis. Aliment Pharmacol Therapeutics 2004;19:729–738.
36 Stamatakos M, Douzinas E, Stefanaki C, Petropoulou C, Arampatzi H, Safioleas C, Giannopoulos G, Chatziconstantinou C, Xiromeritis C, Safioleas M: Ischemic colitis: surging waves of update. Tohoku J Exp Med 2009;218:83–92.
37 El-Assal ON, Besner GE: Heparin-binding epidermal growth factor-like growth factor and intestinal ischemia-reperfusion injury. Semin Pediatr Surg 2004;13:2–10.
38 Franzin G, Grigioni WF, Dina R, Scarpa A, Zamboni G: Mucin secretion and morphological changes of the mucosa in non-neoplastic diseases of the colon. Histopathology 1983;7:707–718.
39 Itoh H, Yagi M, Fushida S, Tani T, Hashimoto T, Shimizu K, Miwa K: Activation of immediate early gene, c-fos, and c-jun in the rat small intestine after ischemia/reperfusion. Transplantation 2000;69:598–604.
40 Mandir N, FitzGerald AJ, Goodlad RA: Differences in the effects of age on intestinal proliferation, crypt fission and apoptosis on the small intestine and the colon of the rat. Int J Exp Pathol 2005;86:125–130.
41 Song J, Wolf SE, Wu XW, Finnerty CC, Gauglitz GG, Herndon DN, Jeschke MG: Starvation-induced proximal gut mucosal atrophy diminished with aging. JPEN J Parenter Enteral Nutr 2009;33:411–416.
42 Xiao ZQ, Moragoda L, Jaszewski R, Hatfield JA, Fligiel SE, Majumdar AP: Aging is associated with increased proliferation and decreased apoptosis in the colonic mucosa. Mech Ageing Dev 2001;122:1849–1864.
43 de la Chapelle A: Genetic predisposition to colorectal cancer. Nat Rev Cancer 2004;4:769–780.
44 Aaltonen LA, Peltomaki P, Leach FS, Sistonen P, Pylkkanen L, Mecklin JP, Jarvinen H, Powell SM, Jen J, Hamilton SR, et al: Clues to the pathogenesis of familial colorectal cancer. Science 1993;260:812–816.
45 Walther A, Johnstone E, Swanton C, Midgley R, Tomlinson I, Kerr D: Genetic prognostic and predictive markers in colorectal cancer. Nat Rev Cancer 2009;9:489–499.
46 Barker N, Ridgway RA, van Es JH, van de Wetering M, Begthel H, van den Born M, Danenberg E, Clarke AR, Sansom OJ, Clevers H: Crypt stem cells as the cells-of-origin of intestinal cancer. Nature 2009;457:608–611.
47 Jiang H, Ju Z, Rudolph KL: Telomere shortening and ageing. Z Gerontol Geriatr 2007;40:314–324.
48 Cawthon RM, Smith KR, O'Brien E, Sivatchenko A, Kerber RA: Association between telomere length in blood and mortality in people aged 60 years or older. Lancet 2003;361:393–395.
49 Lee HW, Blasco MA, Gottlieb GJ, Horner JW 2nd, Greider CW, DePinho RA: Essential role of mouse telomerase in highly proliferative organs. Nature 1998;392:569–574.
50 Rudolph KL, Chang S, Lee HW, Blasco M, Gottlieb GJ, Greider C, DePinho RA: Longevity, stress response, and cancer in aging telomerase-deficient mice. Cell 1999;96:701–712.

51 Flores I, Canela A, Vera E, Tejera A, Cotsarelis G, Blasco MA: The longest telomeres: a general signature of adult stem cell compartments. Genes Dev 2008;22:654–667.

52 Greenberg RA, Allsopp RC, Chin L, Morin GB, DePinho RA: Expression of mouse telomerase reverse transcriptase during development, differentiation and proliferation. Oncogene 1998;16: 1723–1730.

53 Hiyama E, Hiyama K: Telomere and telomerase in stem cells. Br J Cancer 2007;96:1020–1024.

54 Wong KK, Chang S, Weiler SR, Ganesan S, Chaudhuri J, Zhu C, Artandi SE, Rudolph KL, Gottlieb GJ, Chin L, Alt FW, DePinho RA: Telomere dysfunction impairs DNA repair and enhances sensitivity to ionizing radiation. Nat Genet 2000;26:85–88.

55 Rogers LF, Goldstein HM: Roentgen manifestations of radiation injury to the gastrointestinal tract. Gastrointest Radiol 1977;2:281–291.

56 d'Adda di Fagagna F, Reaper PM, Clay-Farrace L, Fiegler H, Carr P, Von Zglinicki T, Saretzki G, Carter NP, Jackson SP: A DNA damage checkpoint response in telomere-initiated senescence. Nature 2003;426:194–198.

57 Shay JW, Wright WE, Werbin H: Defining the molecular mechanisms of human cell immortalization. Biochimica et Biophysica Acta 1991;1072:1–7.

58 Brown JP, Wei W, Sedivy JM: Bypass of senescence after disruption of p21CIP1/WAF1 gene in normal diploid human fibroblasts. Science 1997;277:831–834.

59 Choudhury AR, Ju Z, Djojosubroto MW, Schienke A, Lechel A, Schaetzlein S, Jiang H, Stepczynska A, Wang C, Buer J, Lee HW, von Zglinicki T, Ganser A, Schirmacher P, Nakauchi H, Rudolph KL: Cdkn1a deletion improves stem cell function and lifespan of mice with dysfunctional telomeres without accelerating cancer formation. Nat Genet 2007;39:99–105.

60 Vousden KH, Prives C: Blinded by the light: the growing complexity of p53. Cell 2009;137:413–431.

61 Guo X, Deng Y, Lin Y, Cosme-Blanco W, Chan S, He H, Yuan G, Brown EJ, Chang S: Dysfunctional telomeres activate an ATM-ATR-dependent DNA damage response to suppress tumorigenesis. EMBO J 2007;26:4709–4719.

62 Wong KK, Maser RS, Bachoo RM, Menon J, Carrasco DR, Gu Y, Alt FW, DePinho RA: Telomere dysfunction and Atm deficiency compromises organ homeostasis and accelerates ageing. Nature 2003;421:643–648.

63 Maringele L, Lydall D: EXO1-dependent single-stranded DNA at telomeres activates subsets of DNA damage and spindle checkpoint pathways in budding yeast yku70Delta mutants. Genes Dev 2002;16:1919–1933.

64 Schaetzlein S, Kodandaramireddy NR, Ju Z, Lechel A, Stepczynska A, Lilli DR, Clark AB, Rudolph C, Kuhnel F, Wei K, Schlegelberger B, Schirmacher P, Kunkel TA, Greenberg RA, Edelmann W, Rudolph KL: Exonuclease-1 deletion impairs DNA damage signaling and prolongs lifespan of telomere-dysfunctional mice. Cell 2007;130:863–877.

65 Mimitou EP, Symington LS: Sae2, Exo1 and Sgs1 collaborate in DNA double-strand break processing. Nature 2008; 455:770–774.

66 Celli GB, Denchi EL, de Lange T: Ku70 stimulates fusion of dysfunctional telomeres yet protects chromosome ends from homologous recombination. Nature Cell Biology 2006;8:885–890.

67 Rudolph KL, Hartmann D, Opitz OG: Telomere dysfunction and DNA damage checkpoints in diseases and cancer of the gastrointestinal tract. Gastroenterology 2009;137:754–762.

68 Espejel S, Klatt P, Menissier-de Murcia J, Martin-Caballero J, Flores JM, Taccioli G, de Murcia G, Blasco MA: Impact of telomerase ablation on organismal viability, aging, and tumorigenesis in mice lacking the DNA repair proteins PARP-1, Ku86, or DNA-PKcs. J Cell Biol 2004;167:627–638.

69 Espejel S, Martin M, Klatt P, Martin-Caballero J, Flores JM, Blasco MA: Shorter telomeres, accelerated ageing and increased lymphoma in DNA-PKcs-deficient mice. EMBO Rep 2004; 5:503–509.

70 Maser RS, Wong KK, Sahin E, Xia H, Naylor M, Hedberg HM, Artandi SE, DePinho RA: DNA-dependent protein kinase catalytic subunit is not required for dysfunctional telomere fusion and checkpoint response in the telomerase-deficient mouse. Mol Cell Biol 2007;27: 2253–2265.

71 von Zglinicki T: Role of oxidative stress in telomere length regulation and replicative senescence. Ann NY Acad Sci 2000;908:99–110.

72 Packer L, Fuehr K: Low oxygen concentration extends the lifespan of cultured human diploid cells. Nature 1977;267: 423–425.

73 Lee AC, Fenster BE, Ito H, Takeda K, Bae NS, Hirai T, Yu ZX, Ferrans VJ, Howard BH, Finkel T: Ras proteins induce senescence by altering the intracellular levels of reactive oxygen species. J Biol Chem 1999;274:7936–7940.

74 Balaban RS, Nemoto S, Finkel T: Mitochondria, oxidants, and aging. Cell 2005;120:483–495.

75 Richter C, Park JW, Ames BN: Normal oxidative damage to mitochondrial and nuclear DNA is extensive. Proc Natl Acad Sci USA 1988;85:6465–6467.

76 Shigenaga MK, Hagen TM, Ames BN: Oxidative damage and mitochondrial decay in aging. Proc Natl Acad Sci USA 1994;91:10771–10778.

77 Finkel T, Holbrook NJ: Oxidants, oxidative stress and the biology of ageing. Nature 2000;408:239–247.

78 Adler V, Yin Z, Fuchs SY, Benezra M, Rosario L, Tew KD, Pincus MR, Sardana M, Henderson CJ, Wolf CR, Davis RJ, Ronai Z: Regulation of JNK signaling by GSTp. EMBO J 1999;18:1321–1334.

79 Saitoh M, Nishitoh H, Fujii M, Takeda K, Tobiume K, Sawada Y, Kawabata M, Miyazono K, Ichijo H: Mammalian thioredoxin is a direct inhibitor of apoptosis signal-regulating kinase (ASK) 1. EMBO J 1998;17:2596–2606.

80 Micchelli CA, Perrimon N: Evidence that stem cells reside in the adult *Drosophila* midgut epithelium. Nature 2006;439:475–479.

81 Ohlstein B, Spradling A: The adult *Drosophila* posterior midgut is maintained by pluripotent stem cells. Nature 2006;439:470–474.

82 Lin G, Xu N, Xi R: Paracrine Wingless signalling controls self-renewal of *Drosophila* intestinal stem cells. Nature 2008;455:1119–1123.

83 Fre S, Huyghe M, Mourikis P, Robine S, Louvard D, Artavanis-Tsakonas S: Notch signals control the fate of immature progenitor cells in the intestine. Nature 2005;435:964–968.

84 Ohlstein B, Spradling A: Multipotent *Drosophila* intestinal stem cells specify daughter cell fates by differential notch signaling. Science 2007;315:988–992.

85 Buchon N, Broderick NA, Chakrabarti S, Lemaitre B: Invasive and indigenous microbiota impact intestinal stem cell activity through multiple pathways in *Drosophila*. Genes Dev 2009;23:2333–2344.

86 Wang MC, Bohmann D, Jasper H: JNK signaling confers tolerance to oxidative stress and extends lifespan in *Drosophila*. Dev Cell 2003;5:811–816.

87 Gregorieff A, Clevers H: Wnt signaling in the intestinal epithelium: from endoderm to cancer. Genes Dev 2005;19:877–890.

88 Binnerts ME, Kim KA, Bright JM, Patel SM, Tran K, Zhou M, Leung JM, Liu Y, Lomas WE 3rd, Dixon M, Hazell SA, Wagle M, Nie WS, Tomasevic N, Williams J, Zhan X, Levy MD, Funk WD, Abo A: R-Spondin1 regulates Wnt signaling by inhibiting internalization of LRP6. Proc Natl Acad Sci USA 2007;104:14700–14705.

89 Kim KA, Kakitani M, Zhao J, Oshima T, Tang T, Binnerts M, Liu Y, Boyle B, Park E, Emtage P, Funk WD, Tomizuka K: Mitogenic influence of human R-spondin1 on the intestinal epithelium. Science 2005;309:1256–1259.

90 Ootani A, Li X, Sangiorgi E, Ho QT, Ueno H, Toda S, Sugihara H, Fujimoto K, Weissman IL, Capecchi MR, Kuo CJ: Sustained in vitro intestinal epithelial culture within a Wnt-dependent stem cell niche. Nat Med 2009;15:701–706.

91 Sato T, Vries RG, Snippert HJ, van de Wetering M, Barker N, Stange DE, van Es JH, Abo A, Kujala P, Peters PJ, Clevers H: Single Lgr5 stem cells build crypt-villus structures in vitro without a mesenchymal niche. Nature 2009;459:262–265.

92 Dignass AU, Sturm A: Peptide growth factors in the intestine. Eur J Gastroenterol Hepatol 2001;13:763–770.

93 Haramis AP, Begthel H, van den Born M, van Es J, Jonkheer S, Offerhaus GJ, Clevers H: De novo crypt formation and juvenile polyposis on BMP inhibition in mouse intestine. Science 2004;303:1684–1686.

94 He XC, Zhang J, Tong WG, Tawfik O, Ross J, Scoville DH, Tian Q, Zeng X, He X, Wiedemann LM, Mishina Y, Li L: BMP signaling inhibits intestinal stem cell self-renewal through suppression of Wnt-beta-catenin signaling. Nat Genet 2004;36:1117–1121.

95 Watanabe K, Ueno M, Kamiya D, Nishiyama A, Matsumura M, Wataya T, Takahashi JB, Nishikawa S, Nishikawa S, Muguruma K, Sasai Y: A ROCK inhibitor permits survival of dissociated human embryonic stem cells. Nat Biotechnol 2007;25:681–686.

Tobias Sperka
Institute of Molecular Medicine and
Max Planck Research Group on Stem Cell Aging, University of Ulm
Albert-Einstein-Allee 11
DE–89081 Ulm (Germany)
Tel. +49 731 5036 139, Fax +49 731 5036 102
E-Mail tobias.sperka@uni-ulm.de

Rudolph KL (ed): Molecular Mechanisms of Adult Stem Cell Aging.
Else Kröner-Fresenius Symp. Basel, Karger, 2010, vol 1, pp 79–87

Molecular Mechanisms of Aging in Muscle Stem Cells

Stefan Tümpel

Institute of Molecular Medicine and Max-Planck Research Department on Stem Cell Aging, University of Ulm, Ulm, Germany

Abstract

There is clear evidence of functional decline in muscle tissue during aging. This is reflected in the impaired regenerative potential of aged muscles. It has been suggested that the function of muscle stem cells (satellite cells) may be limited during aging. These satellite cells are required for normal postnatal growth and repair of skeletal muscles in adulthood. During the Else Kröner-Fresenius Symposium on the Molecular Mechanisms of Adult Stem Cell Aging, 3 talks highlighted the role of muscle stem cells during muscle aging. In the first talk, Dr. Thomas A. Rando discussed the changes of histone markers during stem cell activation. The second presentation, by Dr. Thomas Braun, dealt with the role of *Pax7* during muscle growth and the role of Sirtuins during satellite cell aging. The heterogeneity of satellite cells and changes of the transcriptional profile of muscle stem cells during aging was the topic of the third talk, which was presented by Dr. Amy J. Wagers. This chapter will summarize the main points of these talks and in particular focus on epigenetic modifications during satellite cell aging.

Morphological and Molecular Properties of Muscle Stem Cells

Muscle-specific stem cells are called satellite cells, because they appear to orbit the outer surface of the muscle fiber (fig. 1). They are located underneath the basal lamina of mature, differentiated muscle fibers as mitotically quiescent cells. They are activated by injury or muscle growth and then become highly proliferative intermediate progenitor cells.

During this process the satellite cells generate fusion-competent mono-nucleated myoblasts that are capable of fusion and differentiation. Either the myoblasts align and fuse to form multinucleated myotubes to replace the injured muscle or they fuse with existing myofibers to repair damaged muscles [2].

Heterogeneity of Satellite Cells

Several studies have shown variability in the expression of phenotypic markers among satellite cells. One example is the heterogenic expression of Pax3 by satellite cells in different muscles of mice. In most hindlimb muscles and some forelimb and

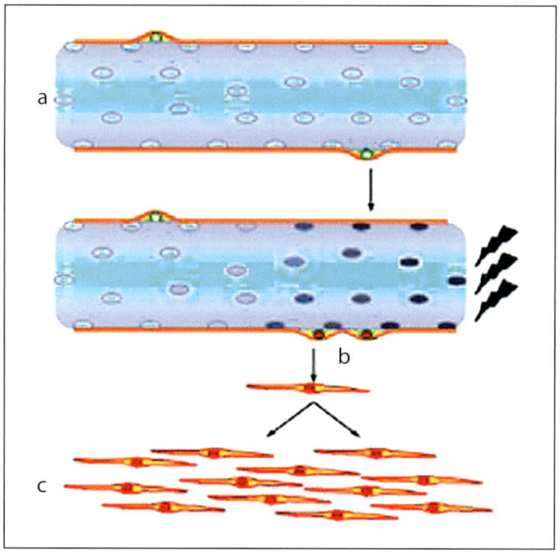

Fig. 1. Schematic diagram of the satellite cell dynamics. The quiescent satellite cells are located beneath the basal lamina (orange) (**a**) and are activated by injury (**b**). The daughter cells migrate from beneath the basal lamina and their progeny proliferate (**c**). Reproduced with permission from Collins et al. [3].

trunk muscles, satellite cells are negative for Pax3, but all these muscles are positive for Pax7. There are some muscles, such as the diaphragm, where the satellite cells show high levels of *Pax3* expression. However, not all the satellite cells express *Pax3* in these muscles [3]. This heterogeneity in the satellite cell population may indicate functional differences or different stages in lineage specification of satellite cells (reviewed in [4]).

At the Else Kröner-Fresenius Symposium on the Molecular Mechanisms of Adult Stem Cell Aging, Dr. Amy J. Wagers discussed the heterogeneity of satellite cells.

Dr. Wagers pointed out that satellite cells consist of both self-renewing stem cells and myogenic precursors [5]. The heterogeneity of satellite cells is evident in the different expression of cell surface markers and differences in proliferation properties. She explained that a unique combination of cell surface markers (CD45$^-$, Sca-1$^-$, Mac-1$^-$, CXCR4$^+$, β1-integrin$^+$, which is abbreviated as CSM4B) identifies a subset of cells within the satellite cell population, which is capable of clonal myogenic differentiation in cell culture assays (fig. 2). After transplantation, they can regenerate damaged muscles and reseed the satellite cell niche [6]. This distinct population was termed skeletal muscle precursors (SMPs).

Origin of Satellite Cells

There are several studies dealing with the origin of the satellite cells. Early studies suggest that satellite cells of the trunk originate from the somites based on quail chick chimeras [7]. In more recent studies, this observation has been confirmed by molecular techniques [8]. Using different transgenic reporter lines, the progenitors of trunk satellite cells have been identified in the central dermomyotome [8]. It has also been shown that most of the limb muscle satellite cells are derived from hypaxial somites [8]. Other resources of satellite cells have been described, including bone marrow and vascular and hematopoietic lineages [9, 10], though the importance of these cells for physiological muscle growth and regeneration remains controversial.

Molecular Basis of Satellite Cell Myogenesis

Satellite cells are dependent on the same regulatory components for determination and differentiation as the myogenic precursor cells in an embryo, including the transcription factors MyoD, Myf5 and Pax7 (fig. 3) [11, 12].

Both *Pax3* and *Pax7* are expressed by quiescent satellite cells [13]. During the activation of satellite cells, *Pax3* and *Pax7* are down-regulated, as the satellite cells start to differentiate [14]. Proportions of activated cells remain undifferentiated, retain *Pax7* expression, and are thought to reconstitute the satellite cell pool. Pax7 seems to inhibit MyoD and myogenin and thereby reacquire a quiescent, undifferentiated state [14]. Both Pax3 and Pax7 simultaneously drive prolif-

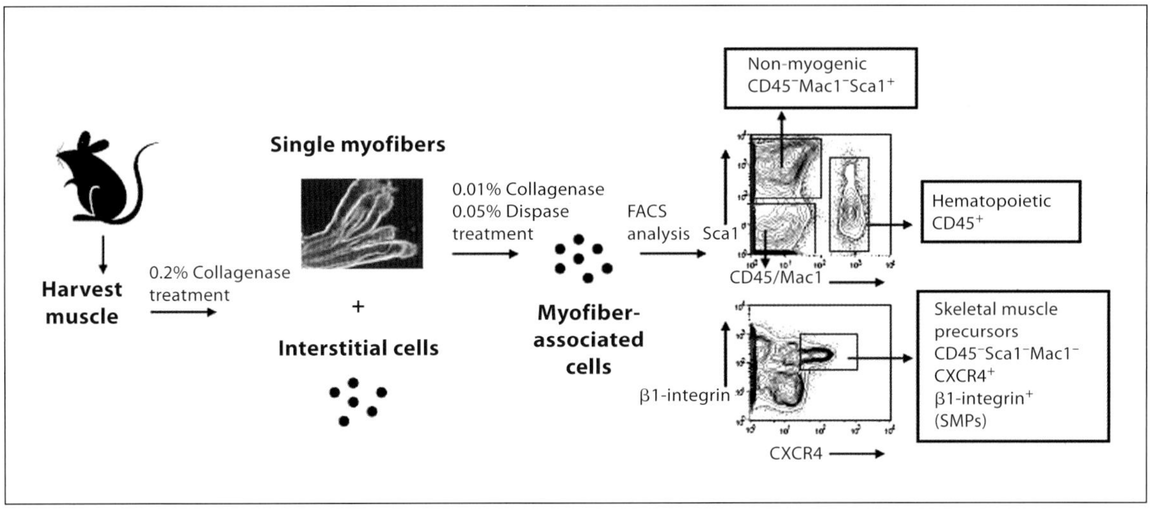

Fig. 2. The experimental procedure for the isolation of skeletal muscle precursors (SMPs). The muscles are harvested and subjected to enzymatic and mechanical treatment. The myofiber-associated cells are then analyzed by FACS. Different subpopulations were identified including non-myogenic cells, hematopoietic cells and the SMPs that showed clonal myogenic potential at single-cell level in the in vitro culture. Reproduced with permission from Cerletti et al. [6].

eration of satellite cell-derived myoblasts and keep them poised for differentiation in response to appropriate environmental cues [15]. *Myf5* is expressed in quiescent cells [16], *MyoD* is expressed as the cells become activated and subsequently differentiate with the expression of *myogenein* [17]. Pax7 has also been shown to have a critical anti-apoptotic function. In the absence of *Pax7*, satellite cells are lost during early postnatal life by apoptosis [13].

Specification of Satellite Cells

Pax7 is a member of the paired-box (Pax) family of transcription factors. It has been proposed that *Pax7* mutant mice are completely devoid of satellite cells and that Pax7 is required for specifying satellite cell lineage [18]. However, Dr. Thomas Braun redefined the function of Pax7 during satellite cell myogenesis and discussed the results during the symposium. *Pax7* knockout mice died because of craniofacial abnormalities. However, a few survived and these were analyzed by the Braun group [19]. The group showed that these mutant mice exhibited postnatal growth and that satellite cells developed normally, although their number was reduced. These data suggest that Pax7 is not required for the specification of myogenic satellite cells.

In order to study the function of Pax7 specifically in satellite cells, the Braun group generated a conditional *Pax7* allele. This strain was used to conditionally ablate the Pax7 function in myogenic cells when crossed with a M*yf5-cre* mouse line. Here, again, the number of satellite cells was initially not reduced, but at a later stage, the number of satellite cells declined. The group further confirmed this observation when muscles were damaged with cardiotoxin in *Pax7*$^{-/-}$ knockout mice. In these mice, regeneration still occured, although its efficiency was decreased. Taken together, these data suggest that Pax7 is not required for the specification of the muscle stem cell lineage, but rather for maintaining and renewing muscle stem cells.

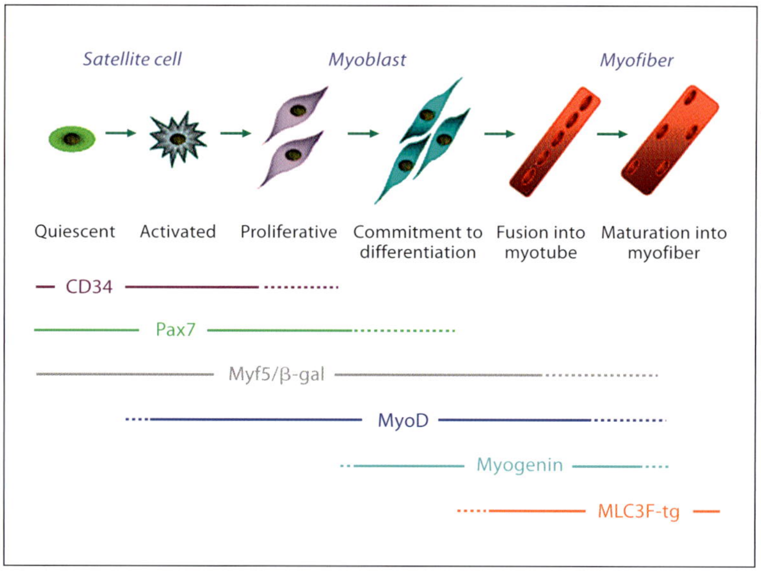

Fig. 3. Temporal expression of different components involved in satellite cell myogenesis. Reproduced with permission from Zammit et al. [1].

Molecular Mechanisms of Satellite Cell Aging

Cell-Intrinsic and -Extrinsic Mechanisms of Satellite Cell Aging

Several reports have shown that extrinsic cues have an impact on satellite cell function during aging, whereas the intrinsic capacity of satellite cells remains largely intact [20–22]. This hypothesis was supported by cell culture experiments using primary mouse satellite cells [20, 22]. When satellite cells of young mice were exposed to serum from old mice, a proportion of the cells lost their myogenic phenotype and acquired a fibroblastic appearance without fusion of the cells. Moreover, *MyoD* and *Pax7* were down-regulated by this treatment [20, 22]. Together, these experiments showed that satellite cells adopted an altered fate by being exposed to systemic environmental factors from old mouse serum. It was then shown in parabiotically paired mice that the systemic environment affects the regeneration potential of satellite cells in vivo. In this experimental system, 2 mice are surgically connected to each other, thereby sharing a circulatory blood system [22]. When a young mouse was attached to an old mouse (heterochronic parabiosis) the regeneration potential of the old muscles increased. This was not the case when an old mouse was attached to another old mouse (isochronic parabiosis). The increased efficiency of regeneration in old satellite cells from heterochronic pairs also paralleled an increased expression of the Notch signaling component *Delta* in the muscle progenitor cells [22].

Wnt Signaling and Its Role in Satellite Cell Aging

Wnt signaling has been shown to be essential during embryogenesis and adult stem cell function [23]. Wnts are involved in embryonic myogenic induction in the paraxial mesoderm and also during muscle fiber differentiation (fig. 4) [24, 25]. Recent work of the group headed by Dr. Thomas A. Rando showed that Wnt signaling has also a critical role in stem cell aging. When adding endogenous Wnt3a to young serum, there was an increased myogenic-to-fibrogenic conversion [20]. If young cells were exposed to old serum together with the Wnt inhibitor Frizzled-related protein 3, the degree of lineage conversion was reduced [20]. Therefore, the effect of the environment is partly mediated by increased Wnt signaling in aged mice. It acts on the quiescent cells, causing them to lose their myogenic lineage and become fibroblasts.

Notch Signaling: Control of Regenerative Competence of Satellite Cells

The role of Notch signaling has been described in several biological processes, including tissue formation during embryonic development [26]. Notch controls developmental patterning, cell fate decisions, and proliferation and maintenance of progenitor cells [27]. The receptor Notch and the ligand Delta are transmembrane proteins. Upon binding of the ligand to the receptor, the intracellular portion of Delta is cleaved, translocated to the nucleus and regulates as a transcription factor the expression of several target genes.

In the last years, it has become apparent that Notch signaling is also activated in postnatal myogenic lineage progression and has a role in satellite cell aging (fig. 4) [28, 29]. In damaged muscles, the Notch pathway is activated and regulates the transition from the quiescent cells to proliferating precursor cells. Ectopic expression of *Notch* promotes proliferation of satellite cells and attenuates the myogenic differentiation of muscle progenitor cells. Conversely, blocking the Notch pathway leads to up-regulation of muscle differentiation genes and reduced proliferation of satellite cells.

Upon injury, the expression of the ligand *Delta* is reduced in satellite cells of aged mice compared to their younger counterparts. The impaired activation of *Delta* is associated with decreased proliferation and expansion of muscle progenitor cells and ineffective regeneration [29]. In addition, there was an inhibition of muscle regeneration when the Notch pathway was inhibited at the site of injury. In contrast, there was improved in vivo regeneration of old muscles when Notch signaling was ectopically induced.

Transcriptional Profile of Muscle Stem Cells during Aging

In her presentation, Dr. Wagers focused on the alterations of muscle stem cells during aging. Using a clonal assay, her group showed that muscle stem cells decline in function and that lost muscle mass is replaced by fat and fibroblasts.

Using microarray, Dr. Wagers' group analyzed the transcriptional profile of muscle stem cells during aging. They observed that the expression profile of the middle-aged SMPs was more similar to that of the old ones than to that of the young ones, despite the fact that their number in middle-aged mice is unaffected. Another observation the Wagers group made was that the signature of the aged SMPs is more similar to that of the progenitor cells in other tissues than to stem cells of other tissues. Together, this lends support to the notion that aged SMPs exhibit further progression along the differentiation pathway.

Epigenetic Regulation of Myogenic Stem Cells during Aging

During the asymmetric division of satellite cells, the 2 daughter cells have 2 different fates: 1 returns to a quiescent state to refill the stem cell pool ($Pax7^+$) and the other will enter the differentiation program ($MyoD^+$ and Myf^+). In satellite cell activation, epigenetic changes are distributed during the division, which permits the coordinated expression of a subset of genes. These chromatin changes coordinate lineage determination by repression and de-repression of lineage-specific genes [30, 31].

Several transcription factors have been shown to be involved in chromatin remodeling during myogenesis including MyoD and Pax7 [32, 33]. It has been shown that MyoD remodels the chromatin of silent loci and thereby activates muscle-specific genes [32]. Pax7 has been shown to enforce satellite cell commitment by chromatin

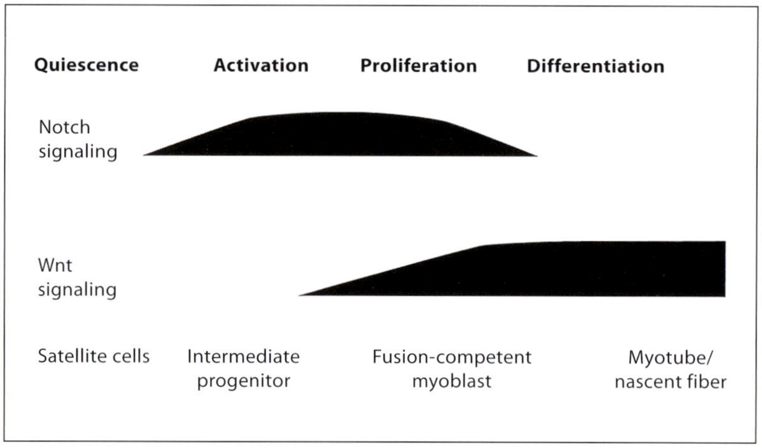

Fig. 4. Schematic diagram on the onset of the different signaling pathways regulating myogenic lineage progression including the Notch, Wnt and GSK3β signaling pathways. Modified from [20].

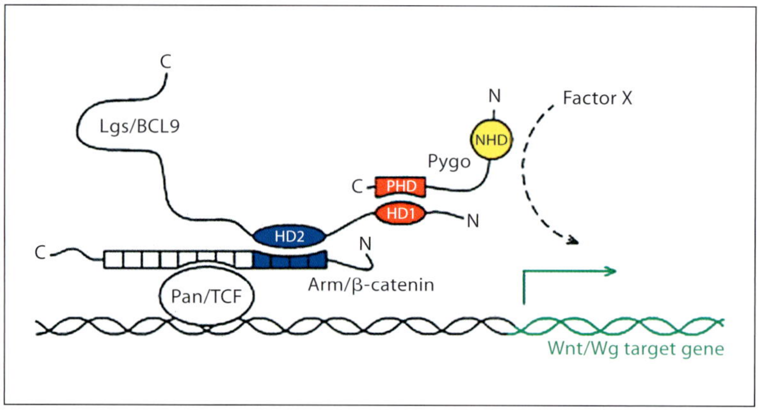

Fig. 5. Schematic diagram showing the components of the quaternary complex composed of BCL9, TCF, armadillo and Pygo. The formation of this complex transduces Wnt signaling by promoting the transcriptional activity of Armadilla. Pygo = Pygopus; Lgs = legless; HD1 = homology domain 1; HD2 = homology domain 2; PHD = plant homology domain; NHD = N-terminal homology domain. Reproduced with permission from Kramps et al. [36].

remodeling. Pax7 is associated with a histone methyltransferase complex that modifies histone H3. This complex binds to the Myf5 locus resulting in methylation of the surrounding chromatin and thereby allowing the activation of gene expression [33].

Link between Wnt Signaling and Epigenetic Modifications

During muscle aging, the proportion of fibrous connective tissue increases [34]. This process could be driven by altered muscle stem cell differentiation. During the meeting, Dr. Rando discussed how the fate conversion from myogenic cells to fibroblasts is regulated. Dr. Rando and coworkers analyzed satellite cells with a microarray approach, focusing on Wnt signaling components. Nineteen putative Wnt target genes were differentially expressed comparing cells in the quiescent and activated states. Interestingly, Foxo binding sites were enriched in a set of down-regulated genes.

The Rando group became interested in one of the Wnt componts, BCL9, which forms a quaternary complex as a cofactor with TCF, Armadilla and Pygo to initiate transcription (fig. 5) [36]. When *BCL9* is absent, a strong wingless phenotype has been described in *Drosophila* [36]. In mammals, there are 2 members of this gene but mutational analysis did not provide any Wnt phenotype, even if both members were disrupted. However, in vitro analysis confirmed that the Wnt pathway is blocked if the function of any of the components of the quaternary complex is impaired.

The Rando group interrupted both murine *BCL9* members by gene targeting and showed that this leads to a reduction in lineage conversion in response to Wnt signaling. This implies that BCL9 is essential for Wnt signaling and also for the age-related phenotype of lineage conversion.

Previously, it has been shown that the BCL9 complex binds to the methylated lysine of histone 3 [35]. Therefore, the complex is not only a transcriptional co-activator, but is also a binding component for the activated genes, which have this epigenetic mark. This would provide a link between Wnt regulator components and epigenetic regulation.

Dr. Rando also discussed the identification of a demethylase, which is a member of the *Jumonji* family. These members have been shown to demethylate lysine on histone 3. Dr. Rando showed that this gene is down-regulated in muscle stem cells upon activation. When it is down-regulated by RNAi, the Rando group observed a dramatic increase in the activation of myogenic stem cells in adult animals. Surprisingly, they did not observe an effect on aged myogenic stem cells.

Together, 2 components that mediate modifications of histone 3 have been identified that are both involved in lineage conversion of muscle progenitor cells. First, the Wnt component BCL9, which forms a complex to methylate lysine of histone 3 and second, a demethylase member of the *Jumonji* family, which leads to demethylation of histone 3.

Role of Sirtuins: Epigenetic Changes during Aging
The Braun group set up a microarray assay to analyze changes of gene expression during aging in mice. Two members of the *Sirtuin* family were on the list of genes with differential expression during aging. Sirtuins comprise a family of protein deacetylases that are regulated by the cofactor NAD^+ and are thereby thought to sense the metabolic state of the organism. They are phylogenetically conserved from bacteria to humans. The yeast Sir2 (silent information regulator 2) is the founding member of the Sirtuins.

So far, 7 mammalian homologues have been identified. They exhibit different sub-cellular localizations. SIRT1, SIRT6 and SIRT7 are localized in the nucleus, whereas SIRT7 is localized in the nucleolus [37]. A group of this gene family has been termed mitochondrial Sirtuins and includes SIRT2, SIRT3, SIRT4 and SIRT5.

They can also be grouped regarding their enzymatic activity [38–40]. On the one hand, SIRT4 and SIRT5 have deacetylase activity, whereas SIRT4 and SIRT6 are mono-ADP-ribosyl transferases. On the other hand, both enzymatic activities have been described for SIRT2 and SIRT3.

In yeast, fly and worm, some of the Sirtuin members have been implicated in generating aging phenotypes [41–43]. In *Caenorhabditis elegans*, it has been shown that increased expression of the SIR2 ortholog, sir-2.1, increases lifespan by up to 50% [43]. Similarly, in *Drosophila* the activation of the SIRT2 ortholog increases lifespan [44].

Sirtuins have been shown to mediate calorie restriction [45], which has been defined as a dietary regimen in which an organism is provided with at least 20% fewer calories than it would naturally consume, while maintaining adequate nutrition [46]. Calorie restriction mediates lifespan extension in several species, including mice [47]. However, the role of the Sirtuin members in mice during aging has yet to be determined.

Conclusions and Perspectives

Our knowledge about aging of satellite cells is growing rapidly. This is underlined by recent talks held during the symposium. In all three talks, similar approaches were used to investigate different aspects of muscle stem cell aging. The analysis of changes in the transcript profiles during aging was discussed by all speakers. Interestingly, these experiments led to the identification of specific epigenetic changes affecting gene transcription during aging. For future studies, one important question is to analyze epigenetic changes at a genome-wide level. These experiments could show to what extent satellite cell aging is affected by chromatin modifications and alterations in gene transcription. It is possible that both cell-extrinsic and -intrinsic mechanisms of satellite cell aging redirect gene transcription by altering the distribution of chromatin-modifying complexes. To reverse age-associated impairments in satellite cell function, it will be important to identify specific causes of muscle progenitor cell aging that are present in the cellular environment as well as intrinsically in cells. In addition, the rescue of cellular functions requires some degree of reversibility of the secondary changes that occur at the genetic and epigenetic levels in response to aging.

References

1 Zammit PS, Partridge TA, Yablonka-Reuveni Z: The skeletal muscle satellite cell: the stem cell that came in from the cold. J Histochem Cytochem 2006;54: 1177–1191.
2 Charge SB, Rudnicki MA: Cellular and molecular regulation of muscle regeneration. Physiol Rev 2004;84:209–238.
3 Collins CA, Olsen I, Zammit PS, Heslop L, Petrie A, Partridge TA, Morgan JE: Stem cell function, self-renewal, and behavioral heterogeneity of cells from the adult muscle satellite cell niche. Cell 2005;122:289–301.
4 Wagers AJ, Conboy IM: Cellular and molecular signatures of muscle regeneration: current concepts and controversies in adult myogenesis. Cell 2005; 122:659–667.
5 Sherwood RI, Christensen JL, Conboy IM, Conboy MJ, Rando TA, Weissman IL, Wagers AJ: Isolation of adult mouse myogenic progenitors: functional heterogeneity of cells within and engrafting skeletal muscle. Cell 2004;119:543–554.
6 Cerletti M, Jurga S, Witczak CA, Hirshman MF, Shadrach JL, Goodyear LJ, Wagers AJ: Highly efficient, functional engraftment of skeletal muscle stem cells in dystrophic muscles. Cell 2008; 134:37–47.
7 Armand O, Boutineau AM, Mauger A, Pautou MP, Kieny M: Origin of satellite cells in avian skeletal muscles. Arch Anat Microsc Morphol Exp 1983;72: 163–181.
8 Schienda J, Engleka KA, Jun S, Hansen MS, Epstein JA, Tabin CJ, Kunkel LM, Kardon G: Somitic origin of limb muscle satellite and side population cells. Proc Natl Acad Sci USA 2006;103:945–950.
9 Asakura A, Rudnicki MA: Side population cells from diverse adult tissues are capable of in vitro hematopoietic differentiation. Exp Hematol 2002;30: 1339–1345.
10 Gussoni E, Soneoka Y, Strickland CD, Buzney EA, Khan MK, Flint AF, Kunkel LM, Mulligan RC: Dystrophin expression in the mdx mouse restored by stem cell transplantation. Nature 1999;401: 390–394.
11 Rudnicki MA, Schnegelsberg PN, Stead RH, Braun T, Arnold HH, Jaenisch R: MyoD or Myf-5 is required for the formation of skeletal muscle. Cell 1993;75: 1351–1359.
12 Maroto M, Reshef R, Munsterberg AE, Koester S, Goulding M, Lassar AB: Ectopic Pax-3 activates MyoD and Myf-5 expression in embryonic mesoderm and neural tissue. Cell 1997;89:139–148.
13 Relaix F, Montarras D, Zaffran S, Gayraud-Morel B, Rocancourt D, Tajbakhsh S, Mansouri A, Cumano A, Buckingham M: Pax3 and Pax7 have distinct and overlapping functions in adult muscle progenitor cells. J Cell Biol 2006;172:91–102.
14 Olguin HC, Olwin BB: Pax-7 up-regulation inhibits myogenesis and cell cycle progression in satellite cells: a potential mechanism for self-renewal. Dev Biol 2004;275:375–388.
15 Collins CA, Gnocchi VF, White RB, Boldrin L, Perez-Ruiz A, Relaix F, Morgan JE, Zammit PS: Integrated functions of Pax3 and Pax7 in the regulation of proliferation, cell size and myogenic differentiation. PLoS One 2009;4:e4475.
16 Beauchamp JR, Heslop L, Yu DS, Tajbakhsh S, Kelly RG, Wernig A, Buckingham ME, Partridge TA, Zammit PS: Expression of CD34 and Myf5 defines the majority of quiescent adult skeletal muscle satellite cells. J Cell Biol 2000; 151:1221–1234.
17 Yablonka-Reuveni Z, Rivera AJ: Temporal expression of regulatory and structural muscle proteins during myogenesis of satellite cells on isolated adult rat fibers. Dev Biol 1994;164:588–603.
18 Seale P, Sabourin LA, Girgis-Gabardo A, Mansouri A, Gruss P, Rudnicki MA: Pax7 is required for the specification of myogenic satellite cells. Cell 2000;102: 777–786.
19 Oustanina S, Hause G, Braun T: Pax7 directs postnatal renewal and propagation of myogenic satellite cells but not their specification. EMBO J 2004;23: 3430–3439.

20 Brack AS, Conboy MJ, Roy S, Lee M, Kuo CJ, Keller C, Rando TA: Increased Wnt signaling during aging alters muscle stem cell fate and increases fibrosis. Science 2007;317:807–810.

21 Shefer G, Van de Mark DP, Richardson JB, Yablonka-Reuveni Z: Satellite-cell pool size does matter: defining the myogenic potency of aging skeletal muscle. Dev Biol 2006;294:50–66.

22 Conboy IM, Conboy MJ, Wagers AJ, Girma ER, Weissman IL, Rando TA: Rejuvenation of aged progenitor cells by exposure to a young systemic environment. Nature 2005;433:760–764.

23 Klaus A, Birchmeier W: Wnt signalling and its impact on development and cancer. Nat Rev Cancer 2008;8:387–398.

24 Anakwe K, Robson L, Hadley J, Buxton P, Church V, Allen S, Hartmann C, Harfe B, Nohno T, Brown AM, Evans DJ, Francis-West P: Wnt signaling regulates myogenic differentiation in the developing avian wing. Development 2003;130:3503–3514.

25 Tajbakhsh S, Borello U, Vivarelli E, Kelly R, Papkoff J, Duprez D, Buckingham M, Cossu G: Differential activation of Myf5 and MyoD by different Wnts in explants of mouse paraxial mesoderm and the later activation of myogenesis in the absence of Myf5. Development 1998;125:4155–4162.

26 Baron M: An overview of the Notch signalling pathway. Semin Cell Dev Biol 2003;14:113–119.

27 Yoon K, Gaiano N: Notch signaling in the mammalian central nervous system: insights from mouse mutants. Nat Neurosci 2005;8:709–715.

28 Conboy IM, Rando TA: The regulation of Notch signaling controls satellite cell activation and cell fate determination in postnatal myogenesis. Dev Cell 2002;3:397–409.

29 Conboy IM, Conboy MJ, Smythe GM, Rando TA: Notch-mediated restoration of regenerative potential to aged muscle. Science 2003;302:1575–1577.

30 Fischle W, Wang Y, Allis CD: Histone and chromatin cross-talk. Curr Opin Cell Biol 2003;15:172–183.

31 Vaquero A, Loyola A, Reinberg D: The constantly changing face of chromatin. Sci Aging Knowledge Environ 2003;2003:RE4.

32 Gerber AN, Klesert TR, Bergstrom DA, Tapscott SJ: Two domains of MyoD mediate transcriptional activation of genes in repressive chromatin: a mechanism for lineage determination in myogenesis. Genes Dev 1997;11:436–450.

33 McKinnell IW, Ishibashi J, Le Grand F, Punch VG, Addicks GC, Greenblatt JF, Dilworth FJ, Rudnicki MA: Pax7 activates myogenic genes by recruitment of a histone methyltransferase complex. Nat Cell Biol 2008;10:77–84.

34 Goldspink G, Fernandes K, Williams PE, Wells DJ: Age-related changes in collagen gene expression in the muscles of mdx dystrophic and normal mice. Neuromuscul Disord 1994;4:183–191.

35 Fiedler M, Sanchez-Barrena MJ, Nekrasov M, Mieszczanek J, Rybin V, Muller J, Evans P, Bienz M: Decoding of methylated histone H3 tail by the Pygo-BCL9 Wnt signaling complex. Mol Cell 2008;30:507–518.

36 Kramps T, Peter O, Brunner E, Nellen D, Froesch B, Chatterjee S, Murone M, Zullig S, Basler K: Wnt/wingless signaling requires BCL9/legless-mediated recruitment of pygopus to the nuclear beta-catenin-TCF complex. Cell 2002;109:47–60.

37 Michishita E, Park JY, Burneskis JM, Barrett JC, Horikawa I: Evolutionarily conserved and nonconserved cellular localizations and functions of human SIRT proteins. Mol Biol Cell 2005;16:4623–4635.

38 North BJ, Marshall BL, Borra MT, Denu JM, Verdin E: The human Sir2 ortholog, SIRT2, is an NAD^+-dependent tubulin deacetylase. Mol Cell 2003;11:437–444.

39 Frye RA: Characterization of five human cDNAs with homology to the yeast SIR2 gene: Sir2-like proteins (sirtuins) metabolize NAD and may have protein ADP-ribosyltransferase activity. Biochem Biophys Res Commun 1999;260:273–279.

40 Shi T, Wang F, Stieren E, Tong Q: SIRT3, a mitochondrial sirtuin deacetylase, regulates mitochondrial function and thermogenesis in brown adipocytes. J Biol Chem 2005;280:13560–13567.

41 Kaeberlein M, McVey M, Guarente L: The SIR2/3/4 complex and SIR2 alone promote longevity in *Saccharomyces cerevisiae* by two different mechanisms. Genes Dev 1999;13:2570–2580.

42 Rogina B, Helfand SL: Sir2 mediates longevity in the fly through a pathway related to calorie restriction. Proc Natl Acad Sci USA 2004;101:15998–16003.

43 Tissenbaum HA, Guarente L: Increased dosage of a sir-2 gene extends lifespan in *Caenorhabditis elegans*. Nature 2001;410:227–230.

44 Wood JG, Rogina B, Lavu S, Howitz K, Helfand SL, Tatar M, Sinclair D: Sirtuin activators mimic caloric restriction and delay ageing in metazoans. Nature 2004;430:686–689.

45 Lin SJ, Defossez PA, Guarente L: Requirement of NAD and SIR2 for life-span extension by calorie restriction in *Saccharomyces cerevisiae*. Science 2000;289:2126–2128.

46 Michan S, Sinclair D: Sirtuins in mammals: insights into their biological function. Biochem J 2007;404:1–13.

47 Weindruch R, Walford RL: Dietary restriction in mice beginning at 1 year of age: effect on life-span and spontaneous cancer incidence. Science 1982;215:1415–1418.

Dr. Stefan Tümpel
Institut für Molekulare Medizin und
Max-Planck-Forschungsgruppe für Stammzellalterung
Universität Ulm
Albert-Einstein-Allee 11
DE–89081 Ulm (Germany)
Tel. +49 731 5036 139, Fax +49 731 5036 102
E-Mail stefan.tuempel@uni-ulm.de

Rudolph KL (ed): Molecular Mechanisms of Adult Stem Cell Aging.
Else Kröner-Fresenius Symp. Basel, Karger, 2010, vol 1, pp 88–94

Free Radicals and Stem Cell Aging

Luis M. Guachalla

Institute of Molecular Medicine and Max Planck Research Group on Stem Cell Aging, University of Ulm, Ulm, Germany

Abstract

For the past 50 years, free radicals have been presumed to be a causal factor of cellular and organismal aging due to their reactive nature. During aging and age-associated disease, there is evidence that free radicals can cause damage to cellular structures. Stem cells and their regenerative potential represent an important therapeutic approach against aging and age-associated diseases. Although in principle stem cells are immortal, there is evidence that their function declines with age. The aging process in stem cells might differ from that of somatic cells, and is unavoidably accompanied by loss or modification of their regenerative capacity. A number of recent papers from different groups uncovered a role of reactive oxygen species in the aging process of stem cells. Here we review the current literature that associates stem cell aging with reactive oxygen species.

Reactive Oxygen Species and Aging

Oxidative stress has been suggested to be one of the causal factors of cellular and organismal aging [1]. In line with this, senescent fibroblasts have accumulated oxidative damage to DNA [2] and proteins [3], and show increased reactive oxygen species (ROS) production [4, 5].

At an organismal level, it has been shown that over-activation of mitochondrial superoxide dismutase *(Sod2)* and catalase targeted to mitochondria prolonged the lifespan of yeast [6, 7] and mice [8], respectively. In addition, deletion of the ROS-producing protein p66shc also significantly extends the lifespan of mice [9]. Moreover, premature aging of diabetic mice is associated with increased oxidative stress and premature senescence in tissues [10]. Thus, it is conceivable that oxidative stress could impair organismal fitness during aging and age-associated disease.

Stem Cell Aging and Oxidative Stress

Stem cell aging is a new and exciting concept that rose concomitantly with the identification of novel multipotent cellular populations. Several lines of evidence indicate that stem cells, although being immortal in principle, can also undergo aging [11–15]. Stem cell aging is proposed to potentially contribute to organismal aging [16], and it is conceivable that the impact of an age-dependent increase in stem cell dysfunction weighs differently in diverse organs (e.g. organs with high vs. low rates of cell turnover). Nonetheless, the mechanisms by which this occurs and its real contribution to human aging remain to be defined.

Oxidative stress is proposed to negatively affect stem cell function following the dogma that free radicals cause damage. However, ROS not only have the potential to corrode cellular structures but also to transduce cellular signals such as those observed during cellular proliferation and differentiation (for review see [17]). Recently, it has been shown that increased ROS levels in *Drosophila* multipotent hematopoietic progenitors can cause precocious differentiation into mature blood cell types [18]. Thus, a tight control of redox states in stem cell compartments is expected to occur in order to maintain the equilibrium between differentiation and self-renewal of stem cells during aging.

Control of ROS Production in Stem Cells

Low Mitochondrial Activity and ROS Levels in Stem Cells: A Role for HIF-1α

The importance of maintaining stability of ROS levels in the stem cell compartments is indicated by work performed in Dr. Toshio Suda's laboratory and presented at the Else Kröner-Fresenius Symposium on the Molecular Mechanisms of Adult Stem Cell Aging. During the meeting, Dr. Suda presented work showing that normal hematopoietic stem cells (HSC) primarily use glycolysis for ATP production rather than the tricarboxylic acid cycle and the OXPHOS (oxidative phosphorylation) system. This shift in the ATP production pathways likely contributes to the maintenance of low ROS levels in long-term HSC. These data stand in agreement with previous studies showing that HSC with high or low cellular and mitochondrial activity exhibit different potentials in repopulation capacity [19–21]. Of note, early HSC with higher self-renewal potential are characterized by lower ROS levels [22], while more differentiated myeloid progenitors have higher ROS levels [23].

Moreover, during the meeting, Dr. Rudolf Wiesner showed that epithelial stem cells can proliferate and differentiate normally into a cornified layer without a functional electron transport chain. These indicates that mitochondrial function, as in the case of HSC, is dispensable for epithelial stem cell maintenance. However, the differentiation of complex structures, such as hair, seems to be severely compromised without a functional electron transport chain.

Additional data presented by Dr. Suda pinpointed that hypoxia-inducible factor-1 α (HIF-1α) is a novel player in regulating stem cell function. HIF-1α is a transcription factor that is activated in cells grown under hypoxic conditions, and which regulates a number of genes involved in energy metabolism, angiogenesis and apoptosis.

New data from Dr. Suda and colleagues indicate that long-term HSC defined by $CD34^{low/-}$ LSK (low or negative expression of CD34, $Sca\text{-}1^+$, $c\text{-}kit^+$, $lineage^-$ cells) maintain lower ROS levels than $CD34^+$ progenitor cells by stabilizing HIF-1α under hypoxic conditions.

Deletion of HIF-1α led to the loss of stem cell quiescence as observed by reduced numbers of HSC in the G_0 phase. Moreover, HSC with the HIF-1α deletion exhibited increased mitochondrial mass and sensitivity to treatment with 5-fluorouracil (a drug used in cancer treatment that is highly toxic to proliferating cells). Together, these data indicate that HIF-1α plays an important role in regulating mitochondrial activity and cellular metabolism in long-term HSC. Importantly, it was shown that stem cell respiratory rates and ROS production are maintained at a minimum by mechanisms still to be investigated.

ATM-Deleted HSC Have Increased Oxidative Stress

Free radicals regulate and are regulated by a number of pivotal factors that determine the cellular outcome in many metabolic processes, including cell proliferation, growth arrest and apoptosis. For example, the Ataxia Telangiectasia Mutated gene *(Atm)* encodes for a kinase that phosphorylates a number of substrates in response to DNA damage. ATM was shown to regulate stem cell self-renewal by controlling ROS levels, and an increase in free radical levels occurred in response to ATM deletion [24]. Accordingly, the link between ATM, oxidative stress and stem cell aging came from studies showing an improvement of stem cell function in $ATM^{-/-}$ mice treated with antioxidants compared to untreated $ATM^{-/-}$ controls [25]. Increased ROS production impairs the functioning of $ATM^{-/-}$ stem cells in part by activating the p38 MAPK pathway, and its inhibition improves repopulating capacity of $ATM^{-/-}$ HSC [25].

FoxO Transcription Factors and Resistance to Oxidative Stress

HSC also control ROS levels through the FoxO transcription factor family. FoxO transcription factors are involved in diverse processes such as cell cycle arrest, DNA repair, cellular differentiation and cell death [26]. FoxO proteins also protect quiescent cells from oxidative stress [27].

Evidence for the role of FoxO transcription factors in modulating resistance to oxidative stress in HSC was revealed in a study by Drs. Tothova, Gilliland, DePinho and colleagues, who conditionally deleted FoxO1, FoxO3 and FoxO4 in the adult hematopoietic system [23]. The multiple FoxO-deficient mice showed abnormal expression of FoxO target genes, leading to increased apoptosis rates and oxidative stress that were translated into defective HSC function. Importantly, reduction of ROS levels in vivo through antioxidant administration (N-acetyl cysteine, NAC) reversed the HSC phenotype of FoxO-deficient mice.

In addition, *Foxo3a* deletion alone was shown to impair HSC function [28]. *Foxo3a*-deleted HSC exhibited increased phosphorylation of p38 MAPK accompanied by elevated ROS levels. Moreover, *FoxO3a*-deficient HSC are defective in maintaining quiescence, and sensitive to myelotoxic injury. Interestingly, as in the case of $ATM^{-/-}$ cells, $FoxO3a^{-/-}$ HSC function was partially rescued with antioxidant treatment [28].

During the symposium, Dr. Ronald A. DePinho presented new data showing that the combined deletion of FoxO genes (1, 3 and 4) also impacts fate decisions in neural stem cells. FoxO-null brain has a precocious aging phenotype with a decline in neural stem cell reserve and function. This phenomenon is partly explained by increased ROS levels, but treatment with NAC, in contrast to the HSC situation, could only partially rescue the defects [29]. Moreover, FoxO-deleted neural stem cells showed dysregulation of peroxiredoxin, gluthatione peroxidase1 and sestrin3, in contrast to FoxO mutant HSC where superoxide dismutases 1 and 3 were deregulated [23, 29]. These data demonstrate that stem cells from different compartments might behave differently when dealing with ROS, at least in the context of FoxO transcription factor deficiency.

ROS Regulation by Bmi1 and HSC Function

Bmi1 is a transcriptional repressor that belongs to the Polycomb group gene family [30]. *Bmi1* was involved in the self-renewal of neuronal, hematopoietic, intestinal and leukemic stem cells [31–34]. In agreement with its function as a p16 repressor, over-expression of Bmi1 extended the replicative lifespan of human fibroblasts [35]. $Bmi1^{-/-}$ mice are born in normal mendelian ratio but die shortly after birth due to decreased hematopoietic progenitors, abnormalities in the skeleton and neurological disorders [30].

During the symposium, Dr. Toren Finkel presented compelling evidence that the decreased lifespan of the *Bmi1* knockout mouse is mediated by increased ROS production. Data from his group demonstrates that thymocytes from $Bmi1^{-/-}$ mice not only exhibit increased ROS production, but also decreased mitochondrial function [36]. Treatment with the oxidant scavenger NAC improved organ morphology and homeostasis in *Bmi1*$^{-/-}$ mice.

In addition, thymocytes from *Bmi1*$^{-/-}$ mice exhibited increased DNA damage as shown by 53BP1 staining (an established marker of DNA damage), and NAC treatment rescued the accumulation of DNA damage foci, indicating that increased ROS levels in response to *Bmi1* deletion induce accumulation of nuclear DNA damage. Moreover, the new study from Dr. Finkel and co-workers demonstrated that deletion of the DNA damage checkpoint protein *Chk2* (a downstream target of ATM activating p53) significantly extended the lifespan of *Bmi1*$^{-/-}$ mice, indicating that the accumulation of nuclear DNA damage and activation of DNA damage checkpoints contribute to the impairment in tissue maintenance in response to *Bmi1* deletion. However, *Chk2* deletion did not rescue the self-renewal defect of *Bmi1*$^{-/-}$ HSC but improved thymopoiesis and early progenitor cell function. Together, these data indicate that the *Chk2*-dependent checkpoint limits progenitor cell function and organ maintenance in response to ROS accumulation, but *Chk2*-independent checkpoints exist at stem cell level limiting stem cell self-renewal in response to increased ROS levels. Together, this study provides an interesting new link between Polycomb genes, mitochondrial function and ROS regulation. In addition, the study points to different checkpoint responses induced by ROS in stem cells compared to progenitor cells.

Downstream Regulators of Oxidative Status in Stem Cells

Although HIF-1α, Bmi1 and FoxO factors regulate expression of a wide range of genes, the data cited above also demonstrate that these factors modulate ROS levels at the stem cell level. Nonetheless, the key downstream effectors that directly modulate free radical production remain elusive.

FoxO transcription factors were shown to upregulate the mitochondrial form of *Sod2* [27]. Although HSC carrying the homozygous or heterozygous deletion for *Sod2* can reconstitute lethally irradiated recipients [37, 38], animals receiving *Sod2*$^{-/-}$ cells are persistently anemic. This anemia is characterized by increased superoxide and hydrogen peroxide production, oxidative damage to proteins and accumulation of iron granules in mature red cells [39]. Thus, although *Sod2* seems to be dispensable in reconstituting lethally irradiated hosts, maintenance of red blood cells requires the expression of SOD2 in order to maintain active red blood cell pools. However, a gene dose reduction in *Sod2*$^{+/-}$ mice was not associated with accelerated aging [40]. Moreover, *Sod2* haplo-insufficiency did not cooperate with telomere dysfunction in limiting stem cell function and organ maintenance in aging mice [38]. Therefore, besides the low production of ROS in stem cells, it seems that the superoxide detoxification system in stem cells is either robust enough or not further affected by changes in *Sod2* expression levels.

Altogether, additional experiments with genetically modified mice in antioxidant/pro-oxidant genes need to be conducted to clarify the role of potential genes directly modulating ROS levels in order to maintain stem cell function. In addition, it needs to be investigated whether alterations in ROS pathways contribute to stem cell aging in wild-type mice and, more importantly, in humans.

Conclusions

There is emerging evidence that a decline in stem cell function can contribute to organismal aging. Although the free radical theory of aging does not hold strong these days, the involvement of free radicals and mitochondria in transducing cellular signals and causing damage in stem and progenitor cell compartments merits further

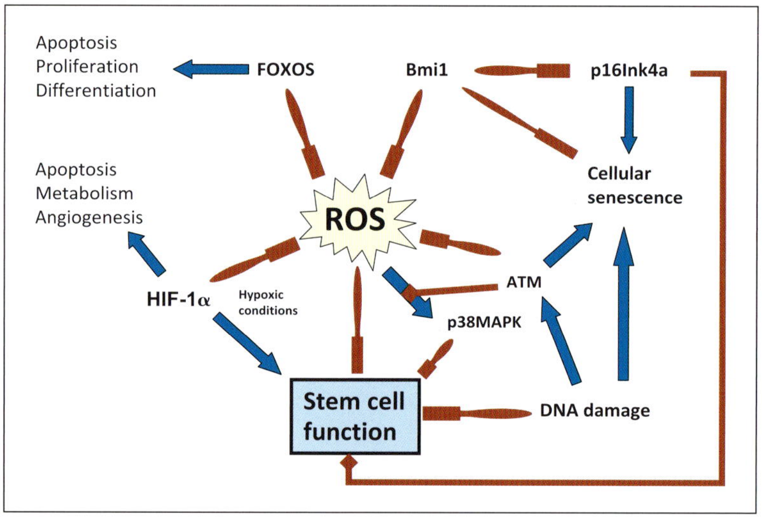

Fig. 1. Factors regulating ROS in stem cells. Stem cells maintain stable ROS levels through a number of factors including ATM, FOXO transcription factors, Bmi1 and, under hypoxic conditions, through HIF-1α. Impairment of stem cell function due to deletion of these factors can be explained by a concomitant elevation of ROS levels. Stem cell mobilization is partially regulated by ROS levels. DNA damage and p16 also impair stem cell function (see chapter 9).

studies. It is also important to note that recent studies have shown that the lifespan-extending effects of calorie restriction might be ROS dependent [41]. Similarly, the beneficial effects of exercise on increasing insulin sensitivity in humans were found to be mediated by exercise-induced ROS [42]. Thus, it is possible that transient peak levels of ROS contribute to an improvement in organismal fitness, whereas the failure in ROS regulation and constantly elevated levels of ROS have negative effects. An important area of future research is also the analysis of ROS toxicity in different cellular compartments. During the current meeting on stem cell aging, experimental evidence for a toxic function of increased ROS levels at the stem and progenitor cell levels were demonstrated (fig. 1). In addition, there is growing evidence that increased ROS levels can contribute to the toxic degeneration of post-mitotic organs such as the brain (for review, see [43]), indicating that post-mitotic cells are also vulnerable to ROS. The contribution of increased ROS levels to the impairment of stem cell function during natural aging remains to be investigated.

References

1 Harman D: Aging: a theory based on free radical and radiation chemistry. J Gerontol 1956;11:298–300.
2 Chen Q, Fischer A, Reagan JD, Yan L, Ames BN: Oxidative DNA damage and senescence of human diploid fibroblast cells. Proc Natl Acad Sci USA 1995;92: 4337–4341.
3 Sitte N, Merker K, von Zglinicki T, Grune T: Protein oxidation and degradation during proliferative senescence of human MRC-5 fibroblasts. Free Radic Biol Med 2000;28:701–708.
4 Passos J, Saretzki G, Ahmed S, Nelson G, Richter T, Peters H, Wappler I, Birket MJ, Harold G, Schaeuble K, Birch-Machin MA, Kirkwood TBL, von Zglinicki T: Mitochondrial dysfunction accounts for the stochastic heterogeneity in telomere-dependent senescence. PLoS Biology 2007;5:e110.
5 Unterluggauer H, Hampel B, Zwerschke W, Jansen-Durr P: Senescence-associated cell death of human endothelial cells: the role of oxidative stress. Exp Gerontol 2003;38:1149–1160.
6 Fabrizio P, Pletcher SD, Minois N, Vaupel JW, Longo VD: Chronological aging-independent replicative life span regulation by Msn2/Msn4 and Sod2 in *Saccharomyces cerevisiae*. FEBS Letters 2004;557:136–142.
7 Fabrizio P, Liou LL, Moy VN, Diaspro A, Selverstone Valentine J, Gralla EB, Longo VD: SOD2 functions downstream of Sch9 to extend longevity in yeast. Genetics 2003;163:35–46.

8 Schriner SE, Linford NJ, Martin GM, Treuting P, Ogburn CE, Emond M, Coskun PE, Ladiges W, Wolf N, Van Remmen H, Wallace DC, Rabinovitch PS: Extension of murine life span by overexpression of catalase targeted to mitochondria. Science 2005;308:1909–1911.
9 Giorgio M, Migliaccio E, Orsini F, Paolucci D, Moroni M, Contursi C, Pelliccia G, Luzi L, Minucci S, Marcaccio M, Pinton P, Rizzuto R, Bernardi P, Paolucci F, Pelicci PG: Electron transfer between cytochrome c and p66shc generates reactive oxygen species that trigger mitochondrial apoptosis. Cell 2005; 122:221–233.
10 Kakoki M, Kizer CM, Yi X, Takahashi N, Kim HS, Bagnell CR, Edgell CJ, Maeda N, Jennette JC, Smithies O: Senescence-associated phenotypes in Akita diabetic mice are enhanced by absence of bradykinin B2 receptors. J Clin Invest 2006;116:1302–1309.
11 Rossi DJ, Bryder D, Zahn JM, Ahlenius H, Sonu R, Wagers AJ, Weissman IL: Cell intrinsic alterations underlie hematopoietic stem cell aging. Proc Natl Acad Sci USA 2005;102:9194–9199.
12 Morrison SJ, Wandycz AM, Akashi K, Globerson A, Weissman IL: The aging of hematopoietic stem cells. Nat Med 1996;2:1011–1016.
13 Sudo K, Ema H, Morita Y, Nakauchi H: Age-associated characteristics of murine hematopoietic stem cells. J Exp Med 2000;192:1273–1280.
14 Chambers SM, Shaw CA, Gatza C, Fisk CJ, Donehower LA, Goodell MA: Aging hematopoietic stem cells decline in function and exhibit epigenetic dysregulation. PLoS Biol 2007;5:e201.
15 Choudhury AR, Ju Z, Djojosubroto MW, Schienke A, Lechel A, Schaetzlein S, Jiang H, Stepczynska A, Wang C, Buer J, Lee HW, von Zglinicki T, Ganser A, Schirmacher P, Nakauchi H, Rudolph KL: Cdkn1a deletion improves stem cell function and lifespan of mice with dysfunctional telomeres without accelerating cancer formation. Nat Genet 2007;39:99–105.
16 Schlessinger D, Van Zant G: Does functional depletion of stem cells drive aging? Mech Ageing Dev 2001;122:1537–1553.
17 Storz P: Reactive oxygen species-mediated mitochondria-to-nucleus signaling: a key to aging and radical-caused diseases. Sci STKE 2006;2006:re3.
18 Owusu-Ansah E, Banerjee U: Reactive oxygen species prime *Drosophila* haematopoietic progenitors for differentiation. Nature 2009;461:537–541.
19 Bertoncello I, Hodgson GS, Bradley TR: Multiparameter analysis of transplantable hemopoietic stem cells. 1. The separation and enrichment of stem cells homing to marrow and spleen on the basis of rhodamine-123 fluorescence. Exp Hematol 1985;13:999–1006.
20 Kim M, Cooper DD, Hayes SF, Spangrude GJ: Rhodamine-123 staining in hematopoietic stem cells of young mice indicates mitochondrial activation rather than dye efflux. Blood 1998;91: 4106–4117.
21 Spangrude GJ, Johnson GR: Resting and activated subsets of mouse multipotent hematopoietic stem cells. Proc Natl Acad Sci USA 1990;87:7433–7437.
22 Jang YY, Sharkis SJ: A low level of reactive oxygen species selects for primitive hematopoietic stem cells that may reside in the low-oxygenic niche. Blood 2007;110:3056–3063.
23 Tothova Z, Kollipara R, Huntly BJ, Lee BH, Castrillon DH, Cullen DE, McDowell EP, Lazo-Kallanian S, Williams IR, Sears C, Armstrong SA, Passegué E, DePinho RA, Gilliland DG: FoxOs are critical mediators of hematopoietic stem cell resistance to physiologic oxidative stress. Cell 2007;128:325–339.
24 Ito K, Hirao A, Arai F, Matsuoka S, Takubo K, Hamaguchi I, Nomiyama K, Hosokawa K, Sakurada K, Nakagata N, Ikeda Y, Mak TW, Suda T: Regulation of oxidative stress by ATM is required for self-renewal of haematopoietic stem cells. Nature 2004;431:997–1002.
25 Ito K, Hirao A, Arai F, Takubo K, Matsuoka S, Miyamoto K, Ohmura M, Naka K, Hosokawa K, Ikeda Y, Suda T: Reactive oxygen species act through p38 MAPK to limit the lifespan of hematopoietic stem cells. Nat Med 2006; 12:446–451.
26 Greer EL, Brunet A: FOXO transcription factors at the interface between longevity and tumor suppression. Oncogene 2005;24:7410–7425.
27 Kops GJPL, Dansen TB, Polderman PE, Saarloos I, Wirtz KWA, Coffer PJ, Huang TT, Bos JL, Medema RH, Burgering BMT: Forkhead transcription factor FOXO3a protects quiescent cells from oxidative stress. Nature 2002;419: 316–321.
28 Miyamoto K, Araki K, Naka K, Arai F, Takubo K, Yamazaki S, Matsuoka S, Miyamoto T, Ito K, Ohmura M, Chen C, Hosokawa K, Nakauchi H, Nakayama K, Nakayama K, Harada M, Motoyama N, Suda T, Hirao A: Foxo3a is essential for maintenance of the hematopoyetic stem cell pool. Cell Stem Cell 2007;1: 101–112.
29 Paik JH, Ding Z, Narurkar R, Ramkissoon S, Muller F, Kamoun WS, Chae SS, Zheng H, Ying H, Mahoney J, Hiller D, Jiang S, Protopopov A, Wong WH, Chin L, Ligon KL, DePinho RA: FoxOs cooperatively regulate diverse pathways governing neural stem cell homeostasis. Cell Stem Cell 2009;5:540–553.
30 van der Lugt NM, Domen J, Linders K, van Roon M, Robanus-Maandag E, te Riele H, van der Valk M, Deschamps J, Sofroniew M, van Lohuizen M: Posterior transformation, neurological abnormalities, and severe hematopoietic defects in mice with a targeted deletion of the bmi-1 proto-oncogene. Genes Dev 1994;8:757–769.
31 Sangiorgi E, Capecchi MR: Bmi1 is expressed in vivo in intestinal stem cells. Nat Genet 2008;40:915–920.
32 Lessard J, Sauvageau G: Bmi-1 determines the proliferative capacity of normal and leukaemic stem cells. Nature 2003;423:255–260.
33 Leung C, Lingbeek M, Shakhova O, Liu J, Tanger E, Saremaslani P, van Lohuizen M, Marino S: Bmi1 is essential for cerebellar development and is overexpressed in human medulloblastomas. Nature 2004;428:337–341.
34 Molofsky AV, Pardal R, Iwashita T, Park IK, Clarke MF, Morrison SJ: Bmi-1 dependence distinguishes neural stem cell self-renewal from progenitor proliferation. Nature 2003;425:962–967.
35 Itahana K, Zou Y, Itahana Y, Martinez JL, Beausejour C, Jacobs JJL, van Lohuizen M, Band V, Campisi J, Dimri GP: Control of the replicative life span of human fibroblasts by p16 and the polycomb protein Bmi-1. Mol Cell Biol 2003;23:389–401.
36 Liu J, Cao L, Chen J, Song S, Lee IH, Quijano C, Liu H, Keyvanfar K, Chen H, Cao LY, Ahn BH, Kumar NG, Rovira II, Xu XL, van Lohuizen M, Motoyama N, Deng CX, Finkel T: Bmi1 regulates mitochondrial function and the DNA damage response pathway. Nature 2009;459:387–392.

37 Friedman JS, Rebel VI, Derby R, Bell K, Huang TT, Kuypers FA, Epstein CJ, Burakoff SJ: Absence of mitochondrial superoxide dismutase results in a murine hemolytic anemia responsive to therapy with a catalytic antioxidant. J Exp Med 2001;193:925–934.
38 Guachalla L, Ju Z, Koziel R, von Figura G, Song Z, Fusser M, Epe B, Jansen-Duerr P, Rudolph KL: *Sod2* haploinsufficiency does not accelerate aging of telomere dysfunctional mice. Aging 2009;1:303–315.
39 Friedman JS, Lopez MF, Fleming MD, Rivera A, Martin FM, Welsh ML, Boyd A, Doctrow SR, Burakoff SJ: SOD2-deficiency anemia: protein oxidation and altered protein expression reveal targets of damage, stress response, and antioxidant responsiveness. Blood 2004;104:2565–2573.
40 Van Remmen H, Ikeno Y, Hamilton M, Pahlavani M, Wolf N, Thorpe SR, Alderson NL, Baynes JW, Epstein CJ, Huang TT, Nelson J, Strong R, Richardson A: Life-long reduction in MnSOD activity results in increased DNA damage and higher incidence of cancer but does not accelerate aging. Physiol Genomics 2003;16:29–37.
41 Schulz TJ, Zarse K, Voigt A, Urban N, Birringer M, Ristow M: Glucose restriction extends *Caenorhabditis elegans* life span by inducing mitochondrial respiration and increasing oxidative stress. Cell Metab 2007;6:280–293.
42 Ristow M, Zarse K, Oberbach A, Kloting N, Birringer M, Kiehntopf M, Stumvoll M, Kahn CR, Bluher M: Antioxidants prevent health-promoting effects of physical exercise in humans. Proc Natl Acad Sci USA 2009;106:8665–8670.
43 Lin MT, Beal MF: Mitochondrial dysfunction and oxidative stress in neurodegenerative diseases. Nature 2006;443:787–795.

Luis M. Guachalla
Institute of Molecular Medicine and
Max Planck Research Group on Stem Cell Aging, University of Ulm
Albert-Einstein-Allee 11
DE–89081 Ulm (Germany)
Tel. +49 731 5036 117, Fax +49 731 5036 102
E-Mail luis.guachalla@gmail.com

Rudolph KL (ed): Molecular Mechanisms of Adult Stem Cell Aging.
Else Kröner-Fresenius Symp. Basel, Karger, 2010, vol 1, pp 95–104

DNA Damage, Checkpoint Responses, and Cell Cycle Control in Aging Stem Cells

Tobias Sperka • Kodandaramireddy Nalapareddy • K. Lenhard Rudolph

Institute of Molecular Medicine and Max Planck Research Group on Stem Cell Aging, University of Ulm, Ulm, Germany

Abstract

Several lines of evidence suggest that the accumulation of DNA damage contributes to organismal aging. Telomere dysfunction, replication errors and toxic metabolites (possibly including reactive oxygen species) represent some of the causes contributing to DNA damage accumulation during aging. DNA damage induces checkpoint responses such as cell cycle arrest or apoptosis. These checkpoints protect the aging organism against cancer but at the same time could contribute to the decline of tissue maintenance and organ function during aging, likely involving an impaired maintenance of functional stem cells. In addition, there is experimental evidence that a tight regulation of cell cycle activity is required to maintain adult stem cells during aging. At the Else Kröner-Fresenius Symposium on the Molecular Mechanisms of Stem Cell Aging, several speakers addressed the role of DNA damage and checkpoint responses in stem cell aging. In this chapter, we review some of the new aspects from the meeting, showing that: (1) stem cells are more resistant to DNA damage-induced checkpoints and exhibit an increased accumulation of DNA damage compared to progenitor cells; (2) deletion of PUMA allows hematopoietic stem cells to survive in response to irradiation-induced DNA damage; (3) p53-dependent checkpoints can protect tissues from age-related atrophy by depleting chromosomally unstable stem cells; (4) Chk2-dependent checkpoints limit progenitor cell function in response to DNA damage induced by reactive oxygen species; (5) genetic modifiers influence the functional role of checkpoint genes in stem cells; (6) FoxO-dependent restriction of cell cycle activity maintains neuronal stem cells, and (7) DNA repair by homologue recombination is required for hematopoietic stem cell expansion and maintenance.

Over recent years, numerous publications have indicated that an accumulation of DNA damage can contribute to organismal aging [1, 2]. In humans, a variety of premature aging syndromes is caused by mutations of genes that contribute to the maintenance of DNA integrity by controlling checkpoint responses, DNA repair or nuclear structure [1]. These syndromes show that the maintenance of DNA integrity is crucial in preventing premature aging. It is conceivable that the rate of DNA damage accumulation contributes to the aging of adult stem cells.

One of the natural causes of DNA damage accumulation in aging cells is the shortening of telomeres, which are specialized nucleoprotein structures that cap the ends of chromosomes. The ends of linear chromosomes need to be distinguished from DNA breaks within the chromosome to avoid inappropriate induction of DNA damage checkpoints and repair responses. Capping of chromosome ends is the main function of telomeres. This allows cells to recognize the intactness of their chromosome ends, thus preventing the induction of DNA damage responses and the development of chromosomal instability. Telomeres shorten with each round of cell division due to the incomplete replication and processing of telomeres during the late S-phase of the cell cycle. Telomere shortening limits the lifespan of primary human cells to 50–70 cell divisions. When telomeres reach a critically short length, they lose the ability to perform their capping function. Uncapped chromosome ends induce DNA damage checkpoints that limit cell proliferation (by induction of senescence) or cell survival (by induction of apoptosis at the crisis stage) [3]. There is growing experimental evidence that the levels of DNA damage, senescence and apoptosis increase in aging tissues [2].

Hematopoietic Stem Cells Are More Resistant to DNA Damage Responses and Accumulate More DNA Damage than Hematopoietic Progenitor Cells

The relative contributions of telomere dysfunction and other forms of DNA damage to natural aging remain to be determined. An important aspect in this regard is the analyses of consequences of DNA damage and telomere dysfunction on stem cell function. Previous studies have shown that DNA damage impairs the function of adult stem cells. Mice with mutations in DNA repair enzymes showed premature aging phenotypes [3, 4]. In addition, telomere dysfunction in late generations of telomerase knockout mice (mTerc$^{-/-}$) has been associated with an impaired maintenance of high-turnover organs and premature aging [5, 6]. At the stem cell level, telomere dysfunction reduced self-renewal and the repopulating capacity of hematopoietic stem cells (HSCs) in serial transplantation experiments [7, 8]. However, transgenic mice with telomerase over-expression did not show increased repopulating capacity of HSCs compared to non-transgenic control mice, although the telomere length was stabilized [9]. Together, these data indicate that telomere-dependent and -independent mechanisms limit HSC self-renewal and repopulating capacity.

During the Else Kröner-Fresenius Symposium on the Molecular Mechanisms of Adult Stem Cell Aging, Dr. Derrick J. Rossi presented new data from his group. Previous work from Dr. Rossi, Dr. Irving L. Weissman and colleagues has shown that cell-intrinsic defects contribute to HSC aging in mice. Particularly, they demonstrated that the age-dependent increase of HSC numbers in C57BL/6J mice and the skewed lineage potential of these aged HSCs (decrease in lymphopoiesis and increase in myelopoiesis) were transplantable and correlated with gene expression changes in HSCs from aged versus young mice [10] (see chapter 2). Recent data from Dr. Weissman, Dr. Rossi and colleagues showed that HSCs exhibit an age-dependent accumulation of DNA damage in mice, which correlated with an impairment of HSC function [4].

Interestingly, Dr. Rossi, Dr. Weissman and colleagues found that the age-dependent accumulation of DNA damage, as measured by γH2AX foci and Comet assay (which measures DNA fragmentation), was increased in HSCs compared to progenitor cells. Dr. Rossi proposed that the low cell cycle activity of HSCs attenuates the responses of HSCs to DNA damage. In agreement with this hypothesis, Dr. Rossi and colleagues found an impaired transcriptional activation of genes involved in DNA damage checkpoints and repair responses in HSCs compared to progenitor cells of aged mice. Dr. Rossi went on to analyze the differences between the reactions of aged and young HSCs to in vitro stimulation of cell cycle activity. Although the vast majority of aged and young HSCs were able to re-enter the cell cycle and gave rise to clonal cultures, a minor subset of aged HSCs showed induction of apoptosis or senescence in response to stimulation. This was not observed in HSCs from young mice. Dr. Rossi proposed a model in which DNA damage accumulates in quiescent HSCs during aging because cell-cycle quiescence inhibits the activation of DNA damage repair and response pathways. However, when stem cells are stimulated to divide, checkpoint responses are activated and DNA damage is largely repaired. When damage cannot be repaired, the cells with high levels of DNA damage are eliminated by apoptosis or cell cycle arrest. In agreement with this hypothesis, previous studies on senescence of human fibroblasts have shown suppressed DNA damage responses in quiescent versus mitogen-stimulated cells [11]. However, it should be noted that initial responses to DNA damage appear to be intact in HSCs as the phosphorylation of H2AX itself represents an upstream response to DNA damage and is readily detectable in quiescent HSCs from old mice.

During the symposium, Dr. Rossi and Dr. Hiromitsu Nakauchi presented data supporting earlier work from Dr. Sean J. Morrison showing that HSCs [lineage marker-negative, c-Kit/Sca-1 double-positive, low or absent levels of CD34 ($CD34^{low/-}$LSK)] can be further purified by the use of the CD150 SLAM marker. Dr. Rossi and Dr. Nakauchi found that this marker distinguished myeloid-biased HSCs (high CD150 expression) from lymphoid-biased HSCs (low CD150 expression). Despite some differences in the experimental set up, both speakers suggested that HSCs that highly express CD150 have a greater self-renewal capacity, possibly explaining the accumulation of myeloid-biased HSCs during aging (see chapter 2). It remains to be investigated whether myeloid-biased HSCs are more resistant to DNA damage or exhibit different rates of DNA damage accumulation.

Inactivation of PUMA Protects Stem Cells in Response to Irradiation-Induced DNA Damage

PUMA is one of the downstream effectors of p53-mediated apoptosis in response to DNA damage [12, 13]. PUMA knockout mice are radio-resistant and, unlike p53 knockout mice, do not show a significant increase in cancer [14]. During the meeting, Dr. Tao Cheng presented new work from his group showing that PUMA deletion improves the maintenance of HSCs in response to γ-irradiation. PUMA was expressed at very low levels in HSCs of non-irradiated mice but was significantly up-regulated in response to irradiation. Together, these data suggest that PUMA-mediated apoptosis depletes HSCs in response to γ-irradiation by inducing apoptosis. However, it remains to be seen whether PUMA only acts at the stem cell level or involves induction of apoptosis at the progenitor cell level.

According to Dr. Cheng's data, PUMA deletion was associated with a significant increase of both short- and long-term survival of γ-irradiated mice. These results were very different from p53 deletion, which confers radio-resistance and improved short-term survival of irradiated mice. However, the long-term survival of irradiated $p53^{-/-}$ mice is limited by rapid tumor development [15, 16]. Dr. Cheng proposed that the intactness of cell cycle arrest checkpoints in $PUMA^{-/-}$ mice might explain their protection from cancer when compared to $p53^{-/-}$ mice that show defects in both cell cycle arrest and apoptosis checkpoints. In agreement with this hypothesis, he found reduced cycling of HSCs in response to γ-irradiation in $PUMA^{-/-}$ mice, whereas HSCs of $p53^{-/-}$ mice continuously entered the cell cycle after γ-irradiation.

p21 is the most prominent downstream target mediating p53-dependent cell cycle arrest in response to DNA damage [17]. p21 has also been shown to mediate cell cycle arrest in senescent human fibroblasts as well as in stem and progenitor cell compartments of aging telomere-dysfunctional mice [18, 19]. Dr. Cheng reported new results on PUMA and p21 double-mutant mice indicating that p21 is only partially responsible for the intactness of cell cycle arrest checkpoints in PUMA$^{-/-}$ mice. His data suggested that additional cell cycle checkpoints in response to DNA damage in HSCs were p21-independent but p53-dependent. While p53$^{-/-}$ HSCs entered multiple rounds of replication in response to irradiation, HSCs from PUMA and p21 double-knockout mice entered at most 1 round of replication in response to γ-irradiation. The nature of these p21-independent checkpoint responses remains to be determined and could involve p53-dependent induction of G_2 cell cycle arrest [20, 21]. In addition, Dr. Cheng presented some evidence indicating that PUMA deletion improved the DNA repair capacities of HSCs in response to γ-irradiation, possibly contributing to cancer prevention in irradiated PUMA$^{-/-}$ mice compared to p53$^{-/-}$ mice.

p53-Dependent Depletion of Chromosomally Unstable Stem Cells Can Protect Tissues from Age-Related Atrophy in the Context of Telomere Dysfunction

The influence of DNA damage checkpoints on the function of adult stem cells was also investigated in a series of studies from Dr. Karl Lenhard Rudolph's laboratory over the past few years. Specifically, the role of DNA damage checkpoints was investigated in the context of telomere dysfunction-induced aging in mTerc$^{-/-}$ mice. Studies from the laboratory of Dr. Ronald A. DePinho had provided the first experimental evidence that telomere dysfunction can limit organ maintenance and induces premature aging in mTerc$^{-/-}$ mice [5, 6]. These defects likely involved the impairment in function of adult tissue stem cells in telomere-dysfunctional mice [7]. In recent work, Dr. Rudolph has shown that the deletion of either Exo1 (exonuclease-1) or p21 can rescue stem cell function and organ maintenance in telomere-dysfunctional mice [19, 22]. Both genes act in the DNA damage signaling pathway and have been shown to limit proliferation of cells in response to telomere dysfunction.

Exo1 is a 5′–3′ exonuclease limiting proliferation of yeast cells in response to telomere dysfunction [23]. It was proposed that processing of dysfunctional telomeres by Exo1 led to generation of single-stranded DNA, allowing the activation of downstream DNA damage signaling. In agreement with this hypothesis, Exo1 deletion rescued the induction of ATR/p53 signaling in telomere-dysfunctional mice. Of note, mTerc and Exo1 double-knockout mice showed a significantly elongated lifespan compared to mTerc$^{-/-}$ Exo1$^{+/+}$ mice. This increase in lifespan was not accompanied by an increase in cancer formation. The induction of chromosomal fusions and genomic instability was shown to increase rates of cancer formation in response to telomere shortening and checkpoint dysfunction [24]. Interestingly, mTerc and Exo1 double-knockout mice did not show an increase in tumor formation, although Exo1 deletion reduced both apoptosis and cell cycle arrest in telomere-dysfunctional mice [22]. A possible explanation for the lack of increased cancer formation in mTerc$^{-/-}$Exo1$^{-/-}$ mice is that the processing of dysfunctional telomeres by Exo1 may contribute to the formation of chromosomal fusions. This would in turn support the evolution of chromosomal instability in response to telomere dysfunction and, hence, tumor formation. Indeed, studies in mTerc$^{-/-}$ Exo1$^{-/-}$ mice revealed a reduced number of chromosomal fusions compared to mTerc$^{-/-}$Exo1$^{+/+}$ mice [22].

Similar to the data on Exo1, the deletion of p21 rescued stem cell function, organ maintenance, and survival of telomere-dysfunctional mice. p21 is the primary target of p53, inducing cell cycle arrest upon sensing senescent human fibroblasts in response to telomere dysfunction [18]. Studies in telomere-dysfunctional mice provided a proof-of-principle that the activation of p21-dependent DNA damage checkpoints impairs the self-renewal and function of HSCs in response to telomere dysfunction. p21 deletion did not lead to an increase in cancer formation in aging telomere-dysfunctional mice. A possible explanation is that p21 left apoptosis checkpoints intact in telomere-dysfunctional mice.

During the meeting, new data were presented on the role of p53 in telomere-driven aging. Previous studies from the DePinho group have shown that p53 deletion can rescue some of the defects in telomere-dysfunctional mice [25]. Specifically, the induction of germ cell apoptosis and infertility was rescued in $p53^{-/-}$ $mTerc^{-/-}$ mice compared to $p53^{+/+}mTerc^{-/-}$ mice. This rescue in germ cell apoptosis resulted in a delayed onset of infertility, which allowed $mTerc^{-/-}p53^{-/-}$ mice to breed for 2 additional generations compared to $mTerc^{-/-}p53^{+/+}$ mice. Similarly, there was a recent report that p53 deletion rescues tissue integrity in the epidermis of mice exposed to acute and high levels of telomere dysfunction induced by deletion of Trf1 (an essential telomere-binding protein for the telomere-capping function) [26]. In contrast to these data on young adult mice, the function of p53 in stem cell aging could not be investigated because $p53^{-/-}$ mice died at a young age from tumor formation (mostly sarcoma and lymphoma) [27, 28]. To avoid this problem, conditional p53 knockout mice were crossed with $mTerc^{-/-}$ mice. In the double-knockout mice, p53 deletion was restricted to the intestinal epithelium as this compartment shows a strong atrophy in response to telomere dysfunction and aging in both humans [29] and mice [19]. Surprisingly, in contrast to the data on p21 knockout mice, the deletion of p53 did not rescue maintenance of the intestinal epithelium in telomere-dysfunctional mice. In fact, $p53^{-/-}mTerc^{-/-}$ mice showed accelerated intestinal atrophy, premature weight loss and a shortened lifespan compared to $p53^{+/+}mTerc^{-/-}$ mice [30]. The study showed that p53 deletion led to an aberrant survival of intestinal stem cells with high rates of DNA damage (γH2AX foci) and chromosomal aberrations (gains and losses). This phenotype did not occur in response to p21 deletion, indicating that p53 has a dual role in intestinal stem cell aging. On one hand, p53 inhibits stem cell maintenance and progenitor cell proliferation by induction of p21. This p21-dependent effect of p53 contributes to the tissue atrophy in response to telomere dysfunction (fig. 1). On the other hand, p53 has a p21-independent function, which leads to the depletion with unstable chromosome stem cells (fig. 1). This effect of p53 protects tissues from aberrant survival of chromosomally unstable stem cells, thus delaying the progressive tissue atrophy in response to telomere dysfunction. According to this model, the deletion of p21 can improve the maintenance and function of stable stem and progenitor cells, resulting in an improved maintenance of telomere-dysfunctional tissues. However, the deletion of p53 leads to an aberrant survival of chromosomally unstable stem cells resulting in aberrant differentiation, p53-independent apoptosis of transient amplifying progenitor cells, and an accelerated development of tissue atrophy in response to telomere dysfunction (fig. 1).

Dr. Eric Brown presented further support for a protective role of p53 in maintaining tissue homeostasis. Previous work from the group has shown that the conditional deletion of ATR in mice leads to depletion of $ATR^{-/-}$ cells in various proliferative tissues of mice. ATR is a DNA damage checkpoint kinase, which is activated following association with resected DNA double-strand and single-strand breaks [31]. Furthermore, ATR has an essential function in maintaining genome stability during S-phase progression by repairing damaged replication forks. Indeed, deletion of ATR causes the rapid accumulation of DNA damage in cycling cells, leading to cell cycle arrest and cell death [32, 33]. However, conditional mosaic deletion of ATR in

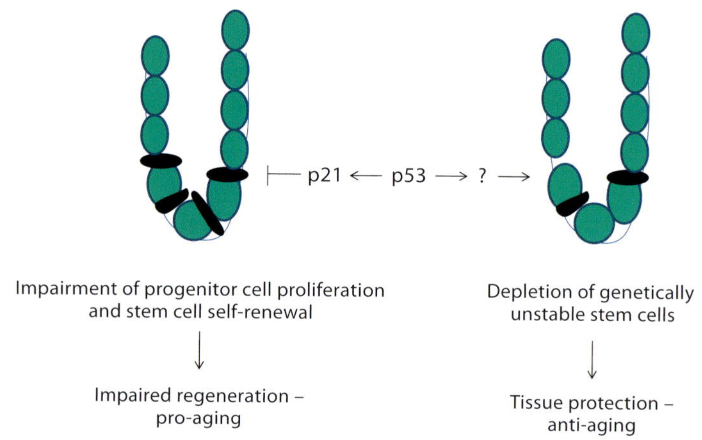

Fig. 1. p53 has a dual role in stem cell aging induced by telomere dysfunction. p53 is activated in response to telomere dysfunction. On one hand, p53 induces the cyclin-dependent kinase inhibitor p21, impairing the proliferation of stem and progenitor cells. This mechanism limits the regenerative reserve during aging and leads to accelerated tissue atrophy in the context of telomere dysfunction [19]. In contrast, p53 induces some unknown target genes that lead to the depletion of chromosomally unstable stem cells. This mechanism has a tissue-protective function in telomere-driven aging by preventing the clonal expansion of unstable stem cells with aberrant differentiation capacity [30].

adult mice is not lethal due to residual subpopulations of ATR-expressing cells that expand to repopulate most tissues following clearance of ATR-deficient cells [34]. Interestingly, ATR mosaic knockout mice (ATR^{mko}) ultimately exhibit age-related phenotypes including hair graying, alopecia, bone loss, premature thymic involution and increased kidney fibrosis, effects that were observed despite prior tissue regeneration by ATR-expressing cells. These prior studies indicated that the additional replication stress induced by the clearance of ATR-deleted cells leads to a premature exhaustion of tissue regenerative potential and premature aging [34]. Such phenotypes may be caused by the limited potential of either the stem or the progenitor cells which directly contribute to tissue maintenance or the accompanying niche cells that control stem cell maintenance and function. In agreement with this hypothesis, the conditional ATR-deficient mice are characterized by reduced abundance of hair follicle bulge stem cells (CD34$^+$), hematopoietic stem and progenitor cells (Lin$^-$Sca$^+$Kit$^+$) and early T lineage progenitors, which correlates with the observed homeostatic deficiencies [34].

In a follow-up project, Dr. Brown and colleagues asked how impeding the clearance of ATR knockout cells might impact the subsequent tissue renewal. To do so, conditional ATR-deficient mice were crossed with p53 knockout mice, a genetic background expected to alleviate many of the cellular effects of DNA damage arising in ATR-deleted cells. However, similar to the studies on telomere-dysfunctional mice (see above), compound p53 deletion accelerated tissue atrophy and delayed compensatory renewal following ATR deletion [35]. In fact, the effects of systemic mosaic deletion of ATR on a p53-deficient background culminated in synthetic lethality in adult mice [35]. The group went on to demonstrate that the absence of p53 prolonged the survival of genetically damaged ATR$^{-/-}$ stem and progenitor cells in the skin (CD34$^+$) and significantly impaired the repopulation of affected organs by ATR-expressing cells. Together, these results support a new concept indicating that the p53 checkpoint prevents the accumulation of damaged cells in tissues, which increases the efficiency and fidelity of tissue regeneration from healthy stem and progenitor cells.

Chk2-Dependent Checkpoints Limit Progenitor Cell Function in Response to DNA Damage Induced by ROS

During the symposium, new data from Dr. Toren Finkel's group demonstrated the first evidence that Bmi1 executes a functional role in maintaining mitochondrial function and controlling levels of reactive oxygen species (ROS). Bmi1 is so far mainly known as a Polycomb group (PcG) transcriptional repressor inhibiting expression of the tumor-suppressive *Ink4a/Arf* locus [36]. Previous studies found that the deletion of Bmi1 de-repressed the expression of p16Ink4a and p19Arf, and led to a marked reduction of HSC maintenance by inhibiting their self-renewal [37, 38]. Deletion of Bmi1 has now been shown to alter mitochondrial function, leading to a significant increase in ROS and an accumulation of nuclear DNA damage in hemato-lymphopoietic cells. This process provoked a DNA damage response and subsequently depletion of thymocytes and hematopoietic stem cells (see chapter 2) [39]. The DNA damage response was characterized by 53BP1 DNA damage foci formation and activation of Chk2. 53BP1 is necessary for the G_2/M and intra-S-phase checkpoint setup and is involved in Chk2 and p53 induction [40]. Chk2 in turn acts as a DNA damage checkpoint kinase contributing to the activation of p53 in response to DNA damage [41–43]. Interestingly, the work showed that a systemic anti-oxidant treatment (eliminating the cause of DNA damage) or deletion of Chk2 (eliminating the response to DNA damage) partially rescued hemato-lymphopoietic defects and the survival of $Bmi1^{-/-}$ mice independently of the up-regulation of Ink4a/Arf. Of note, the data from the group of Dr. Finkel showed that Chk2 deletion on a $Bmi1^{-/-}$ background rescued hemato-lymphopoiesis without rescuing stem cell maintenance and self-renewal, as measured in competitive transplantation experiments. Together, these data indicate that Chk2 induces DNA damage checkpoints in response to ROS at the progenitor cell level but likely not at the stem cell level. Additionally, Dr. Finkel's work provides evidence for the coordinated regulation of 2 independent pathways linked to stem and progenitor cell function by the polycomb transcriptional repressor Bmi1, namely cell cycle control through Ink4a/Arf and cellular metabolism via mitochondria and ROS regulation.

Genetic Modifiers Influence the Functional Role of Checkpoint Genes in Stem Cells

Together, the above data indicate that DNA damage and checkpoint responses have profound effects on the maintenance of stem cell function and organismal aging. During the meeting, Prof. Gerald de Haan presented elegant new work from his laboratory demonstrating that genetic modifiers influence the functional role of the p21 cell cycle arrest checkpoint on maintaining the function of HSCs. Previous work from Dr. Cheng, Dr. David Scadden and colleagues had demonstrated that p21 has an important role in maintaining HSC quiescence [44]. The study has shown that in mice of a mixed background (129 and C57BL/6J) deletion of p21 led to increased cycling of HSCs and early exhaustion of HSCs under stress conditions. In contrast, Prof. de Haan had found that p21 deletion did not affect the maintenance of HSCs quiescence and function in mice of the C57BL/6J strain [45]. During the symposium, Prof. de Haan presented new data comparing the role of p21 deletion in mice of the C57BL/6J and DAB strains. Interestingly, he reproduced the previous data from Dr. Cheng and Dr. Scadden by showing a critical role of p21 for stem cell maintenance in DAB mice. On this background, p21 deletion had a profound effect on HSC sensitivity to replicative stress and $p21^{-/-}$ HSCs showed a strongly reduced repopulation capacity compared to $p21^{+/+}$ control HSCs. However, in agreement with Prof. de Haan's previous data, deletion of p21 had no measurable effect on HSC function in

C57BL/6J mice. Together, these data demonstrate that genetic modifiers can strongly affect the role of specific checkpoint genes in stem cell aging. These findings will have strong implications for translating results from knockout mouse studies on checkpoint genes and stem cell function towards human aging. The data indicate that genetic modifiers need to be evaluated for both the understanding of stem cell biology as well as for the possible targeting of stem cells in novel therapies.

FoxO-Dependent Restriction of Cell Cycle Maintains Neuronal Stem Cells in Mouse Brain

The data on p21 deletion in some strains of mice have shown that the maintenance of quiescence is an important regulatory mechanism to maintain functional stem cells during aging. Similar results have also been reported for the maintenance of neuronal stem cells in the brain [46]. During the meeting, Dr. DePinho presented new data showing that the restriction of cell cycle activity in neuronal stem cells is also controlled by FoxO transcription factors [47]. Specifically, FoxO-1, -3 and -4 triple-knockout mice showed an increased brain size and increased proliferation of neuronal stem cells in the neuronal sphere culture assay. However, Dr. DePinho and coworkers recognized a strong decrease in neuronal stem cell numbers in aging mouse brains correlating with neuronal atrophy. The group identified several downstream targets of the FoxOs in the brain including ASPM1 (a gene that causes microcephaly in human brains when mutated) and SOST1 (a negative regulator of the canonical Wnt signaling pathway). The latter observation was interesting since previous studies have shown that an over-activation of Wnt can result in premature depletion of stem cells in various tissue compartments [48–50]. Together, Dr. DePinho proposed that FoxO is an important regulator of stem cell homeostasis during aging by fine-tuning the balance of stem cell quiescence, proliferation and differentiation.

Conclusion

In conclusion, several speakers at the Else Kröner-Fresenius Symposium on the Molecular Mechanisms of Stem Cell Aging highlighted the importance of DNA integrity and cell cycle control for the maintenance of functional stem cells during aging. Interestingly, the activation of DNA damage responses can have beneficial or negative effects on the stem cell maintenance and function. It will be an important direction of future work to delineate the influence of DNA damage and checkpoint responses on basic pathways controlling stem cell self-renewal and differentiation. Moreover, the influence of DNA damage and checkpoint responses on the systemic environment and the stem cell niche have to be considered as important extrinsic factors contributing to the decline in stem cell function during aging.

References

1 Lieber MR, Karanjawala ZE: Ageing, repetitive genomes and DNA damage. Nat Rev 2004;5:69–75.

2 Nalapareddy K, Jiang H, Guachalla Gutierrez LM, Rudolph KL: Determining the influence of telomere dysfunction and DNA damage on stem and progenitor cell aging: what markers can we use? Exp Gerontol 2008;43: 998–1004.

3 Nijnik A, Woodbine L, Marchetti C, Dawson S, Lambe T, Liu C, Rodrigues NP, Crockford TL, Cabuy E, Vindigni A, Enver T, Bell JI, Slijepcevic P, Goodnow CC, Jeggo PA, Cornall RJ: DNA repair is limiting for haematopoietic stem cells during ageing. Nature 2007; 447:686–690.

4 Rossi DJ, Bryder D, Seita J, Nussenzweig A, Hoeijmakers J, Weissman IL: Deficiencies in DNA damage repair limit the function of haematopoietic stem cells with age. Nature 2007;447: 725–729.

5 Lee HW, Blasco MA, Gottlieb GJ, Horner JW 2nd, Greider CW, DePinho RA: Essential role of mouse telomerase in highly proliferative organs. Nature 1998;392:569–574.
6 Rudolph KL, Chang S, Lee HW, Blasco M, Gottlieb GJ, Greider C, DePinho RA: Longevity, stress response, and cancer in aging telomerase-deficient mice. Cell 1999;96:701–712.
7 Allsopp RC, Morin GB, DePinho R, Harley CB, Weissman IL: Telomerase is required to slow telomere shortening and extend replicative lifespan of HSCs during serial transplantation. Blood 2003;102:517–520.
8 Samper E, Fernandez P, Eguia R, Martin-Rivera L, Bernad A, Blasco MA, Aracil M: Long-term repopulating ability of telomerase-deficient murine hematopoietic stem cells. Blood 2002;99: 2767–2775.
9 Allsopp RC, Morin GB, Horner JW, DePinho R, Harley CB, Weissman IL: Effect of TERT over-expression on the long-term transplantation capacity of hematopoietic stem cells. Nat Med 2003;9:369–371.
10 Rossi DJ, Bryder D, Zahn JM, Ahlenius H, Sonu R, Wagers AJ, Weissman IL: Cell intrinsic alterations underlie hematopoietic stem cell aging. Proc Natl Acad Sci USA 2005;102:9194–9195.
11 Satyanarayana A, Greenberg RA, Schaetzlein S, Buer J, Masutomi K, Hahn WC, Zimmermann S, Martens U, Manns MP, Rudolph KL: Mitogen stimulation cooperates with telomere shortening to activate DNA damage responses and senescence signaling. Mol Cell Biol 2004;24:5459–5474.
12 Jeffers JR, Parganas E, Lee Y, Yang C, Wang J, Brennan J, MacLean KH, Han J, Chittenden T, Ihle JN, McKinnon PJ, Cleveland JL, Zambetti GP: Puma is an essential mediator of p53-dependent and -independent apoptotic pathways. Cancer Cell 2003;4:321–328.
13 Villunger A, Michalak EM, Coultas L, Mullauer F, Bock G, Ausserlechner MJ, Adams JM, Strasser A: p53- and drug-induced apoptotic responses mediated by BH3-only proteins puma and noxa. Science 2003;302:1036–1038.
14 Michalak EM, Villunger A, Adams JM, Strasser A: In several cell types tumour suppressor p53 induces apoptosis largely via Puma but Noxa can contribute. Cell Death and Differ 2008;15:1019–1029.
15 Kemp CJ, Wheldon T, Balmain A: p53-deficient mice are extremely susceptible to radiation-induced tumorigenesis. Nat Genet 1994;8:66–69.
16 Lee JM, Abrahamson JL, Kandel R, Donehower LA, Bernstein A: Susceptibility to radiation-carcinogenesis and accumulation of chromosomal breakage in p53 deficient mice. Oncogene 1994;9:3731–3736.
17 Deng C, Zhang P, Harper JW, Elledge SJ, Leder P: Mice lacking p21CIP1/WAF1 undergo normal development, but are defective in G1 checkpoint control. Cell 1995;82:675–684.
18 Brown JP, Wei W, Sedivy JM: Bypass of senescence after disruption of p21CIP1/WAF1 gene in normal diploid human fibroblasts. Science 1997;277:831–854.
19 Choudhury AR, Ju Z, Djojosubroto MW, Schienke A, Lechel A, Schaetzlein S, Jiang H, Stepczynska A, Wang C, Buer J, Lee HW, von Zglinicki T, Ganser A, Schirmacher P, Nakauchi H, Rudolph KL: Cdkn1a deletion improves stem cell function and lifespan of mice with dysfunctional telomeres without accelerating cancer formation. Nat Genet 2007;39:99–105.
20 Cross SM, Sanchez CA, Morgan CA, Schimke MK, Ramel S, Idzerda RL, Raskind WH, Reid BJ: A p53-dependent mouse spindle checkpoint. Science 1995;267:1353–1356.
21 Fukasawa K, Choi T, Kuriyama R, Rulong S, Vande Woude GF: Abnormal centrosome amplification in the absence of p53. Science 1996;271:1744–1747.
22 Schaetzlein S, Kodandaramireddy NR, Ju Z, Lechel A, Stepczynska A, Lilli DR, Clark AB, Rudolph C, Kuhnel F, Wei K, Schlegelberger B, Schirmacher P, Kunkel TA, Greenberg RA, Edelmann W, Rudolph KL: Exonuclease-1 deletion impairs DNA damage signaling and prolongs lifespan of telomere-dysfunctional mice. Cell 2007;130:863–877.
23 Maringele L, Lydall D: EXO1-dependent single-stranded DNA at telomeres activates subsets of DNA damage and spindle checkpoint pathways in budding yeast yku70Delta mutants. Genes Dev 2002;16:1919–1933.
24 Artandi SE, Chang S, Lee SL, Alson S, Gottlieb GJ, Chin L, DePinho RA: Telomere dysfunction promotes non-reciprocal translocations and epithelial cancers in mice. Nature 2000;406:641–645.
25 Chin L, Artandi SE, Shen Q, Tam A, Lee SL, Gottlieb GJ, Greider CW, DePinho RA: p53 deficiency rescues the adverse effects of telomere loss and cooperates with telomere dysfunction to accelerate carcinogenesis. Cell 1999; 97:527–538.
26 Martinez P, Thanasoula M, Munoz P, Liao C, Tejera A, McNees C, Flores JM, Fernandez-Capetillo O, Tarsounas M, Blasco MA: Increased telomere fragility and fusions resulting from TRF1 deficiency lead to degenerative pathologies and increased cancer in mice. Genes Dev 2009;23:2060–2075.
27 Donehower LA, Harvey M, Slagle BL, McArthur MJ, Montgomery CA Jr, Butel JS, Bradley A: Mice deficient for p53 are developmentally normal but susceptible to spontaneous tumours. Nature 1992;356:215–221.
28 Jacks T, Remington L, Williams BO, Schmitt EM, Halachmi S, Bronson RT, Weinberg RA: Tumor spectrum analysis in p53-mutant mice. Curr Biol 1994; 4:1–7.
29 Vulliamy T, Dokal I: Dyskeratosis congenita. Semin Hematol 2006;43:157–166.
30 Begus-Nahrmann Y, Lechel A, Obenauf AC, Nalapareddy K, Peit E, Hoffmann E, Schlaudraff F, Liss B, Schirmacher P, Kestler H, Danenberg E, Barker N, Clevers H, Speicher MR, Rudolph KL: p53 deletion impairs clearance of chromosomal instable stem cells compromising integrity of the intestine in aging telomere dysfunctional mice. Nat Genet 2009;41:1138–1143.
31 Cimprich KA: Probing ATR activation with model DNA templates. Cell Cycle 2007;6:2348–2354.
32 Brown EJ, Baltimore D: ATR disruption leads to chromosomal fragmentation and early embryonic lethality. Genes Dev 2000;14:397–402.
33 Brown EJ, Baltimore D: Essential and dispensable roles of ATR in cell cycle arrest and genome maintenance. Genes Dev 2003;17:615–628.
34 Ruzankina Y, Pinzon-Guzman C, Asare A, Ong T, Pontano L, Cotsarelis G, Zediak VP, Velez M, Bhandoola A, Brown EJ: Deletion of the developmentally essential gene ATR in adult mice leads to age-related phenotypes and stem cell loss. Cell stem cell 2007;1:113–126.

35 Ruzankina Y, Schoppy DW, Asare A, Clark CE, Vonderheide RH, Brown EJ: Tissue regenerative delays and synthetic lethality in adult mice after combined deletion of Atr and Trp53. Nat Genet 2009;41:1144–1149.
36 Jacobs JJ, Kieboom K, Marino S, DePinho RA, van Lohuizen M: The oncogene and Polycomb-group gene bmi-1 regulates cell proliferation and senescence through the ink4a locus. Nature 1999; 397:164–168.
37 Lessard J, Sauvageau G: Bmi-1 determines the proliferative capacity of normal and leukaemic stem cells. Nature 2003;423:255–260.
38 Park IK, Qian D, Kiel M, Becker MW, Pihalja M, Weissman IL, Morrison SJ, Clarke MF: Bmi-1 is required for maintenance of adult self-renewing haematopoietic stem cells. Nature 2003;423: 302–305.
39 Liu J, Cao L, Chen J, Song S, Lee IH, Quijano C, Liu H, Keyvanfar K, Chen H, Cao LY, Ahn BH, Kumar NG, Rovira, II, Xu XL, van Lohuizen M, Motoyama N, Deng CX, Finkel T: Bmi1 regulates mitochondrial function and the DNA damage response pathway. Nature 2009;459:387–392.
40 Wang B, Matsuoka S, Carpenter PB, Elledge SJ: 53BP1, a mediator of the DNA damage checkpoint. Science 2002; 298:1435–1438.
41 Chehab NH, Malikzay A, Appel M, Halazonetis TD: Chk2/hCds1 functions as a DNA damage checkpoint in G(1) by stabilizing p53. Genes Dev 2000;14: 278–288.
42 Hirao A, Kong YY, Matsuoka S, Wakeham A, Ruland J, Yoshida H, Liu D, Elledge SJ, Mak TW: DNA damage-induced activation of p53 by the checkpoint kinase Chk2. Science 2000;287: 1824–1827.
43 Shieh SY, Ahn J, Tamai K, Taya Y, Prives C: The human homologs of checkpoint kinases Chk1 and Cds1 (Chk2) phosphorylate p53 at multiple DNA damage-inducible sites. Genes Dev 2000;14:289–300.
44 Cheng T, Rodrigues N, Shen H, Yang Y, Dombkowski D, Sykes M, Scadden DT: Hematopoietic stem cell quiescence maintained by p21cip1/waf1. Science 2000;287:1804–1808.
45 van Os R, Kamminga LM, Ausema A, Bystrykh LV, Draijer DP, van Pelt K, Dontje B, de Haan G: A Limited role for p21Cip1/Waf1 in maintaining normal hematopoietic stem cell functioning. Stem Cells 2007;25:836–843.
46 Kippin TE, Martens DJ, van der Kooy D: p21 loss compromises the relative quiescence of forebrain stem cell proliferation leading to exhaustion of their proliferation capacity. Genes Dev 2005; 19:756–767.
47 Paik JH, Ding Z, Narurkar R, Ramkissoon S, Muller F, Kamoun WS, Chae SS, Zheng H, Ying H, Mahoney J, Hiller D, Jiang S, Protopopov A, Wong WH, Chin L, Ligon KL, DePinho RA: FoxOs cooperatively regulate diverse pathways governing neural stem cell homeostasis. Cell Stem Cell 2009;5:540–553.
48 Kirstetter P, Anderson K, Porse BT, Jacobsen SE, Nerlov C: Activation of the canonical Wnt pathway leads to loss of hematopoietic stem cell repopulation and multilineage differentiation block. Nature Immunol 2006;7:1048–1056.
49 Liu H, Fergusson MM, Castilho RM, Liu J, Cao L, Chen J, Malide D, Rovira, II, Schimel D, Kuo CJ, Gutkind JS, Hwang PM, Finkel T: Augmented Wnt signaling in a mammalian model of accelerated aging. Science 2007;317: 803–806.
50 Scheller M, Huelsken J, Rosenbauer F, Taketo MM, Birchmeier W, Tenen DG, Leutz A: Hematopoietic stem cell and multilineage defects generated by constitutive beta-catenin activation. Nature Immunol 2006;7:1037–1047.

K. Lenhard Rudolph
Institute of Molecular Medicine and
Max Planck Research Group on Stem Cell Aging, University of Ulm
DE–89081 Ulm (Germany)
Tel. +49 731 50 36 100, Fax +49 731 50 36 102
E-Mail lenhard.rudolph@uni-ulm.de

Subject Index